The Emergence of Qatar

The Emergence of Qatar

The Turbulent Years 1627-1916

H. Rahman

KEGAN PAUL
London • New York

First published in 2005 by
Kegan Paul Limited
UK: P.O. Box 256, London WC1B 3SW, England
Tel: 020 7580 5511 Fax: 020 7436 0899
E-Mail: books@keganpaul.com
Internet: http://www.keganpaul.com
USA: 61 West 62nd Street, New York, NY 10023
Tel: (212) 459 0600 Fax: (212) 459 3678
Internet: http://www.columbia.edu/cu/cup
BAHRAIN: bahrain@keganpaul.com

Distributed by:
Marston Book Services Ltd
PO Box 269
Abingdon
Oxfordshire OX14 4YN
Tel: +44 (0) 1235 465500, Fax: +44 (0) 1235 465555
Email: direct.order@marston.co.uk

Columbia University Press
61 West 62nd Street, New York, NY 10023
Tel: (212) 459 0600 Fax: (212) 459 3678
Internet: http://www.columbia.edu/cu/cup

ISBN: 0-7103-1213-X

British Library Cataloguing in Publication Data

Library of Congress Cataloging-in-Publication Data
Applied for.

Contents

Contents

Contents

Contents

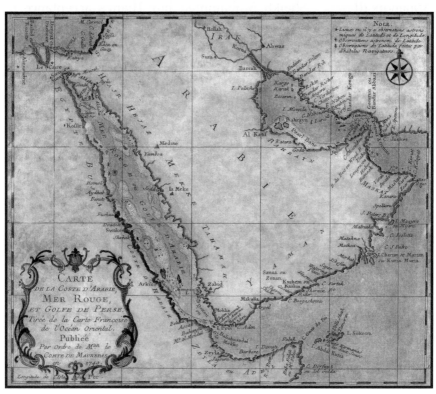

Map 1: Arabian Coast showing Qatar as 'Katara',
Compte de Maurepas, 1740

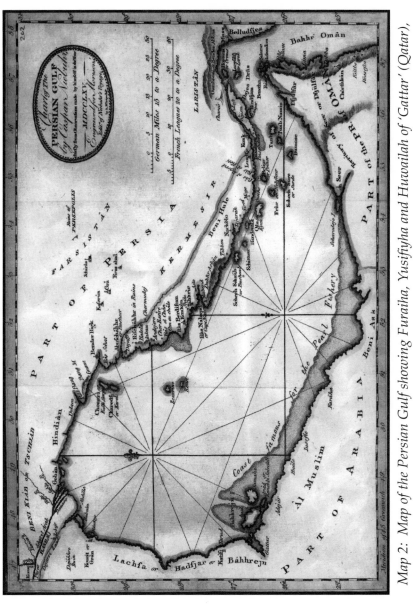

Map 2: Map of the Persian Gulf showing Furaiha, Yusifiyha and Huwailah of 'Gattar' (Qatar), Carsten Niebuhr, 1765

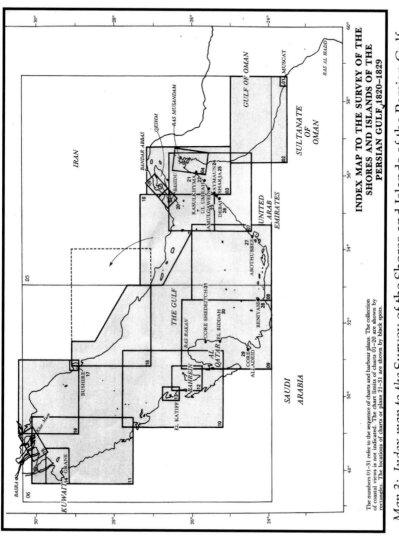

The numbers 01–31 refer to the sequence of charts and harbour plans. The collection of coastal views is not indicated. The chart limits of charts 01–20 are shown by rectangles. The locations of charts or plans 21–31 are shown by black spots.

INDEX MAP TO THE SURVEY OF THE
SHORES AND ISLANDS OF THE
PERSIAN GULF, 1820–1829

Map 3: Index map to the Survey of the Shores and Islands of the Persian Gulf by J. M. Guy and G. B. Brucks, 1820–1829

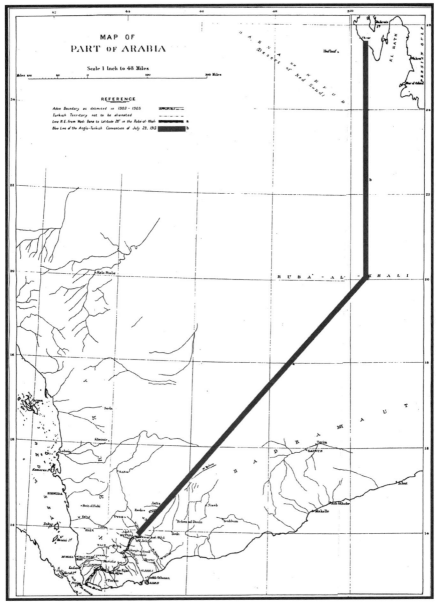

Map 4: Line dividing the Territory of Qatar and Arabia,
Anglo-Ottoman Convention of 29th July 1913

List of Plates

Glossary

Amir	Ruler
Baghla	Traditional ocean-going vessel of the Arabian Gulf and up to 135 feet long
Battil	Coastal trading vessel, also used in warfare
Bazaar	Market
Dhow	War vessel
Ghuree	Defence tower
Imam	Leader
Jalibut	Cargo and pearling boat 6 and about 50 feet in length
Khor	Back water or shallow
Majlis	Council, Parliament
Mudir	Local chief
Mutessarrif	A Government official below the governor
Naib	Deputy
Naqib	Chief or leader
Naqib al Ashraf	Leader of the descendants of the Prophet Mohammad (pbh) in a particular locality
Nahiye	Ottoman local government division
Pasha	Ottoman Official title given to ministers or provincial governors

Qada	Ottoman local government division
Qadi	Judge appointed by the State with administrative and judicial functions
Qaim maqam	Deputy, a high official, Chief Minister or a Provincial governor
Qalat	Fort
Sanjak	Ottoman administrative division
Shaikh	Title given to members of the ruling families of the Arabian Gulf states, to the chief or head of an Arab tribe, family or village and to religious leader
Souq	Market
Sultan	Ruler
Wali	Governor
Vizier	Minister
Zakat	Religious tax

Abbreviations

ADM	Admiralty
BOA	Basbakanlik Osmanli Arsivi
CAB	Cabinet
CID	Committee of Imperial Defence
FO	Foreign Office
FS	Foreign Secretary
GG in C	Governor General in Council
G of I	Government of India
HCS	Honorary Company's Service
HMG	His / Her Majesty's Government
HMS	His / Her Majesty's Service
IO	India Office
IOR	India Office Records
RS	Indian Rupees
L/P&S	Political and Secret Department Records
MS	Manuscript

NCCAH	National Council for Culture, Arts & Heritage
SGB	Secretary to the Government of Bombay
SGI/FD	Secretary to the Government of India in the Foreign Department
S/S for FA	Secretary of State for Foreign Affairs
S/S for India	Secretary of State for India
PG	Persian Gulf
PRO	Public Record Office
TL	Turkish Lira
US/S	Under Secretary of State
V in C	Viceroy in Council

PREFACE

The history of Qatar from the Portuguese bombardment of 1627 to the conclusion of the Treaty of 1916 is a hitherto untold story of destructions, wars, battles, conflicts, intrigues, conspiracy and strategic contests originating in the ashes of the north-west coast of the peninsula and brought to a conclusion at al-Bida (later Doha). On 3 November 1916, the peninsula of Qatar became a sovereign and independent state, the foundation for which was laid by Shaikh Mohammad bin Thani in the early 1850s. The peninsula had emerged as a distinct political entity as a result of the Anglo-Qatari Treaty of 1868, but it was the Anglo-Ottoman convention of 1913, reflecting the two powers' strategic interests in the Gulf in general and Qatar in particular, which defined the status of Qatar for the first time, after many years of obscurity over the issue. Despite its unique geographical location and the Gulf pearl belt off its shores, Qatar was the last of the Arabian Gulf littoral countries to emerge as an independent political entity.

Four main factors were responsible for the comparatively late emergence of Qatar: the absence of a strong and unified national leadership until the rise of the al-Thani; the collapse of the 1868 Treaty owing to the Ottoman occupation of al-Bida; the prolonged Ottoman occupation in the face of growing discontent among the Qataris and the challenge posed by Imperial Britain; and finally and most importantly, the indecision of the British Gulf policy makers and their diametrically opposed views on Qatar until the opening of the formal Anglo-Qatari negotiations in Doha, which resulted in the emergence of an independent Qatar under British limited protection.

No book has yet examined the years of frustration and upheaval that led to the emergence of Qatar. The present work attempts, in an authoritative and comprehensive fashion, to do just that. John Gordon Lorimer, the eminent British civil servant and complier of the Persian Gulf Gazetteer, was the pioneer historian of Qatar, but his study was brief and confined to the years from 1766 to 1907, and is a mere 48 pages long.[1] Lorimer's references to the major events and factors that contributed to the making of Qatar are sketchy. Few works on the earlier history of Qatar have appeared over the last two decades in either Arabic or English.[2] No studies have examined in detail the circumstances that led to the shift of power from Zubara to al-Bida, which eventually emerged as the cockpit of regional rivalries as well as those of the international powers. Neither Lorimer nor any other scholar has shed light on the circumstances under which Shaikh Mohammad bin Thani emerged as the founder of the al-Thani dynasty in Qatar. No author has fairly appraised Shaikh Jasim's vision of Qatar throughout his career. The origin of Qatar's struggle for British protection have likewise eluded in-depth analysis.

This book attempts to fill these lacunae, casting new light on all the untapped issues of Qatar's history. It explores and discovers the rich past history of Qatar and civilization for the first time. This book provides a detailed chronological account of the major factors and events leading to the emergence of Qatar as a separate entity in 1868 and its slow and steady march towards independence; it sets this account against the background of growing Anglo-Ottoman strategic contests in Qatar, leading to the conclusion of the 1913 convention, which cleared the way for the signing of the historic 1916 Treaty. It will also serve as a guide and reference book for students, scholars, historians, Gulf policy makers and travellers interested in the history of Qatar, and the Gulf in general.

This book departs from previous research on Qatar by making extensive reference to unpublished British official correspondence preserved in the British Library and the Public Record Office (now the National Archives) in London. Ottoman official records on Qatar, covering the period from 1871 to 1913 and some Arabic sources have also been consulted. Over two decades of working in the Research & Studies Department of the Office (Diwan Amiri) of His Highness the Amir of the State of Qatar, together with my extensive research on Qatar and its relations with the neighbouring states and on the thorny boundary issues in the region, have furnished me with knowledge of the unexplored periods of Qatar's history and enabled me to set this study in a broader perspective. Visits to almost all the historical sites and key locations of Qatar and long association with high ranking Qatari officials, as well as frequent attendance at the leading Qataris' traditional Majlis[3] – a private social forum or gathering of people from different backgrounds – has deepened my understanding of the history of Qatar, its culture, heritage, customs and traditions. Notes.

1.   See 'History of Qatar' in John Gordon Lorimer, Gazetteer of the Persian Gulf, Oman and Central Arabia, Historical, Vol. I, Hants, UK: Gregg International Publications, Ltd., 1970, pp. 787-835.

2.   Mustafa Murad al-Dabagh, *Qatar-Madiha Wa Hadiruha*, Beirut: Dar al-Taliha, 1961. Abdul Aziz al-Mansour, *Al-Tattawur al-Siyasi Li Qatar Fi al-Fatrah Ma Bayan 1868-1016*, Kuwait: Chains Publications for Printing, Publishing and Distribution, 1980. Mohammed Sharif al-Shibani, *Emarat Qatar al-Arabiya Bayan al Madhi Wa al Hadhir*, Beirut: Culture House, 1962. Frederick F. Anscombe, *The Ottoman Gulf: The Creation of Kuwait, Saudi Arabia and Qatar*, New York: Columbia University Press, 1997. Zekeriya Kursun, *The Ottomans in Qatar: A History of Anglo-Ottoman Conflicts in the Persian Gulf*, Istanbul: The Isis Press, 2002. Rosemarie Said Zahlan, *The Creation of Qatar*, London: Croom Helm Ltd., 1979.

3.   The Majlis of His Highness the Amir of the State of Qatar differs from that of the private Majlis. In fact, this traditional Majlis of H H the Amir is an informal parliament and decision making body and provides H H the Amir with an opportunity to listen to the day to day problems of the people and their views on the efficiency of the government services and works.

Acknowledgements

The publication of this book would not have been possible without the constant help and encouragement I have received from my friends, colleagues and well-wishers. I am grateful to my mentor Dr. M. E. Yapp, Emeritus Professor of Modern Near & Middle Eastern History at the University of London, for reading the manuscript and making valuable comments. I am also grateful to Dr. Shaikh Hasan bin Mohammed bin Ali al-Thani, Chairman, Center for Arabian Gulf Historical Research and Cultural Advisor for the Qatar Foundation, Doha for his help in publishing the book. Thanks are due to HE Mohammad Esa al-Mohannadi former Minister of Cabinet Affairs, Sate of Qatar, HE Mohammad Esa al-Mohannadi, former Minister of Cabinet Affairs, State of Qatar, HE Abdullah Gorab al-Marri, Director of Research & Studies Dept., Diwan Amiri, State of Qatar for his advice and encouragement. Dr. Hassan Saleh al-Ansari, Professor of Qatar University, Mr. Mohammad Khalifa al-Attiyah, former Head of Research and Studies Division, Diwan Amiri, Mr. Hossain Rajab al-Ismail, former Director of Museums and Antiquities Dept., NCCAH, State of Qatar, Mr. Mohammed Said al-Baluchi, Director, Dept. of Museums and Antiquities, NCCAH, Mr. Sami Barzak and Mr. El Noor Osman of Diwan Amiri also deserve thanks for their encouragement. Special thanks and appreciation go to two colleagues of many years standing: Mr. Sultan al-Jaber, Head of Historical Research & Documents Division and Mr. Ahamed al-Muftah, chief Librarian, NCCAH, for their valuable suggestions and for helping me translate the Arabic sources. Thanks are also due to Mr. Mohammed Afsar, Archive Unit, NCCAH, for assisting me in research and Mr. Alex Skinner for his editing of the book. I would like to acknowledge the kind assistance of Mr. Nasser al-Jassem of NCCAH, Mr. Mohammed Hammam Fikhri, Center for

Acknowledgements

Arabian Gulf Historical Research, the staff of the National Archives, and the Institute of Historical Research, London . Finally, I wish to thank my publisher Mr. Peter Hopkins for his interest and enthusiasm in my work and for being this book's first fan. The hard work and inspiration for this book has to be credited to the encouragement and support I have received from my wife, Ghanima, my three daughters, Nadia, Heba and Sarah, and my son, Saad. All unpublished Crown copyright documents from IOR and PRO appear by permission of the Controller of HM Stationary Office. The opinions expressed in the book are entirely my own.

CHAPTER ONE

Historical and Geo-Political Setting

'Katara' or 'Guttur' or 'El-Katr' and now Qatar is a peninsula, which juts into the southern Arabian Gulf almost midway between the mouth of the Shatt al-Arab waterway to the north-west and the strait of Hormuz (to the north of Oman at the Gulf's eastern end). The breadth of the Gulf between Ras Rakkan, at the tip of the Qatar peninsula and the Persian coast at Ras Mustaf was estimated at 100 miles in the first edition of the Persian Gulf Pilot.[1] Qatar, with its arid desert climate, extends the length of the peninsula, which is about 120 miles long and 60 miles wide and covers a total area of 7110 square miles. In medieval times the name 'Bahrayn' was applied to the whole Arabian littoral from the vicinity of Kuwait to Qatar including the islands of Bahrain. Qatar lay on the border of the two major geographical divisions into which eastern Arabia had been split, Bahrain and Oman.

QATAR IN ANCIENT AND MEDIEVAL TIMES

Qatar played a vital role in the development of maritime trade and commerce even in the Stone Age and Neolithic period. Archaeological discoveries of pottery, flint, flint scraper tools and the rim of a painted ceramic vessel with part of the flange from al-Ubaid, Mesopotamia, at Bir Zekrit and Ras Abaruk, situated on the western coast of Qatar, indicates Ras Abaruk's trade and commercial links with the Mesopotamian civilization, which flourished in the land between the Tigris and the Euphrates (in present-day Iraq) during the $5^{th} - 4^{th}$ millennium BC. Ras Abaruk continued to grow in importance and it became a fishing station where foreign vessels used to land to dry fish in 140 BC. Further settlements and a trading point developed in and around Umm al-Ma, situated to the north-west of Ras Abaruk, during the reign of the Sasanids in Persia (3rd century AD – 6th century AD). Ras Abaruk and Umm al-Ma, played a conspicuous role in the Sasanid commerce, particularly in the shipment of precious pearls and purple dye.[2]

Islam spread throughout the entire Arabian Gulf region in the seventh century. During the Islamic period settlements continued to grow and a system of administration developed on the west coast of Qatar peninsula. In the year 628, Al Mundhir bin Sawa al-Tamimi, the ruler of Qatar at the time, along with his subjects, embraced Islam when the Prophet Mohammad's (Peace be upon Him) emissary al Ala al-Hadrami visited al-Tamimi.[3] Eventually, trade and commercial links were established between Qatar and Arabia. Historians have failed to identify al-Tamimi's seat of administration, but it may have been at either Umm al-Ma or Murwab in the Joghbi area. This theory is lent supported by the archaeological discovery of about 100 small stone-built houses and the fortified palaces of a tribal leader at Murwab, which date to the early Islamic period. The Murwab palace or fort was built on the ruins of an earlier fort destroyed by fire. Furthermore, an early Islamic village site with rectangular wall lines similar to those at Murwab was unearthed one kilometer north of al-Nama, south-west of Zubara.[4]

During the reign of the Umayyads (661-750 AD), and the Abbasids (750-1258 AD) in Damascus and Baghdad respectively, trade and commerce grew in Qatar. It became a famous camel and horse breeding centre during the Umayyad period, while the Abbasid period witnessed the development of the pearling industry in the rich waters around Qatar. The demand for Qatari pearl also increased in the East, as far away as China. As mercantile activities intensified on the western coast of Qatar, settlements began to spring up there, particularly at Murwab, between Zubara and Umm al-Ma, where, as mentioned above, 100 or more small stone-built houses were constructed.[5]

TOWNS AND TOPOGRAPY IN MODERN TIMES

The development of modern settlements is best understood in the historical and geographical context. It is interesting to note that Carsten Niebuhr's map of the Persian Gulf, published in 1765, failed to show Qatar as a peninsula. Niebuhr's map is confusing and not accurate. Although he visited Arabia in 1764, he prepared his map without visiting Qatar personally and relied on local Arab informants as well as English sea captains. Niebuhr's Qatar included 'Guttur' [al-Bida], 'Adsjar' (probably al-Khor), al-Huwailah, Yusufiyah and Furaiha.[6] In 1809, more than four decades after the publication of Niebuhr's map, Captain John Wainwright referred to the Qatar coast as 'the coast of danger' because of its inaccessibility from the sea, an opinion based on his voyage to the Gulf.[7] Major Colebrook's report on the Persian Gulf littoral of 10 September 1820, gives the first description of the major towns of Qatar:

> Zobarah [Zubara] .... protected by a tower and occupied at present merely for the security of fishermen that frequent it. It has a Khoor with three fathoms water which Buggalahs may enter.
>
> Khoor Hassam – Before the seat of Aula Yelaina [Jalahima] has a Khoor admitting Buggulahs drawing 2½ fathoms. It has no inhabitant at present.
>
> Limel [Jumail] .... at one time could furnish 1,000 men (5,000 of the tribe Maheude [al-Maadhid] and 500 Bokuara [Bu Kawara] a branch of the Beni Tamem) defended by a square Ghurry on the seaside. Fresh water abundant.
>
> Foreihel [Furaiha] – Deserted formerly occupied by the Bu Sumaitt.
>
> Ul Yusvee [al Yusufiyah] and Roes [Ruwais]- Are two villages to the west of Ras Rekkan [Rakkan], the inhabitants of which had latterly removed to Bahrein. Limel, Yusvee and Roes, have Khoors or back waters, which are shallow. Batils can enter at high tide, but at low water they are quite dry, and the sea coast very shallow. Yusvee has a square Ghurry and fresh water.
>
> Phoerol [Fuwairet] – To the East of Ras (cape) Reckan, the inhabitants removed to Bahrein, has no Khoor the coast on this side the cape, is bolder and may be approached by vessels within gunshot.
>
> Howeleh [al-Huwailah] – Is defended by a square Ghurry, containing good water, and is frequented by fishermen in the season. It was inhabited by a remnant of the once powerful tribe of Musellim [al-Musallam], now incorporated with the Utubis. The tranquillity restored to the Gulf,

will probably cause all these places to be reoccupied, the coast being well watered, which is not the cases in Bahrein.

Guttur- Or Ul Buddee [al-Bida] once a considerable town, is protected by two square Ghurries, near the sea shore; but containing no fresh water they are incapable of defence except against sudden incursions of Bedouins, another Ghurry is situated two miles inland and has fresh water with it. This could contain 200 men. There are remaining at Ul Buddee about 250 men, but the original inhabitants, who may be expected to return from Bahrein, will augment them to 900 or 1,000 men, and if the Doasir tribe, who frequent the place as divers, again settle in it, from 600 to 800 men.[8]

The major tribes living in north-west Qatar at that time were: al-Bu Kuwara, al-Maadhid, al-Bu Sumait and al-Musallam, the last of which were 'incorporated with the Utubis', as Colebrook mentioned. All these coastal towns and villages were situated near the Arabian Gulf pearl banks and the inhabitants had been practising pearl fishing since time immemorial.

THE BRITISH SURVEY OF QATAR

The first systematic survey of the topography of the Gulf was carried out between 1820 and 1829 by the Officers of the East India Company's Bombay Marine. In 1821 Lieutenant J. N. Guy surveyed from the Trucial Coast to the tip of the Qatar peninsula at Ras Rakkan; Captain George Barnes Brucks then continued up to the Shatt al-Arab. The task was completed in 1825 and the results were published in 1829. Describing briefly the topography of Khor al-Odaid channel, Brucks mentioned Wakra, which was composed of coral and small rocks and situated about ten miles east of al-Bida. Although Brucks' report made no reference to any settlement at Wakra, he examined al-Bida in detail:

Al Bidder Town is in lat. 25° 16' 50" N., long. 51° 33' E. It is situated at the bottom of a harbour, formed by the reefs before mentioned. The entrance is only half a mile wide, having three and four fathoms water in it in mid channel, and rather less towards the northern reef, and towards the southern two fathoms. As the shoals show themselves distinctly, to enter the harbour you must be guided by them, keeping rather nearer to the northern one. The narrow part is not more than half a mile in length. When through, steer for the town, and anchor abreast the eastern tower, in three and a half fathoms, about half a mile off shore. This place contains about four hundred Arabs of the Nahune [Naim]; Dooasir [Dawasir], and Abookara [al Bu Kawarah] Tribes, and is frequented by the Monasir and other wandering tribes. In the pearl season the inhabitants are augmented to about twelve hundred, it being convenient to the banks, and so completely sheltered. The people are mostly fishermen; they have one or two trading boats, and, like all other inhabitants of this coast, take a large share in the pearl fishery.[9]

The total population of al-Bida was therefore four hundred, consisting of the Naim, al Bu Kuwara, Dawasir tribes. There were also two or three towns and a 'Ghuree' in al-Bida for defensive purposes.[10]

Therefore, by the 1820s a township had developed with an impressive population, containing several tribes and featuring growing commerce. Al-Huwailah, another major town, to the north of the peninsula, also drew Brucks' attention. This coastal town, older than al-Bida, was defended by a small Ghuree and was inhabited by about four hundred and fifty of the al Bu Kuwara tribe. The people were engaged in maritime trade and during the pearl fishing season al-Huwailah was one of the principal trade centres. In addition to these towns, Brucks surveyed almost all the coastal towns and villages on the coast of Qatar including al Khor, Fuwairet, Ruwais, Ras Rakkan, Abu Dhaluf, Jumail, Yusufiyah, Khor Hassan, Furaiha and Zubara as well as Rubaiya. The total population of these places varied from fifty to one hundred. In the absence of any central authority these coastal towns and villages were governed by the local Shaikhs. Brucks' last point of survey on the Qatar coast was the Warden islands (later Hawar islands), situated about one mile off the coast of Qatar. Two fishing villages owned by the migratory Dawasir tribe were found on the main island.[11]

The successful completion of the survey was a remarkable achievement by Brucks. The survey included comprehensive information regarding the tribes, towns, villages and resources of the Arabian shores. However, as incorrect longitudinal information appeared in the survey, Lieutenant C. G. Constable and A. W. Stiffe were employed in 1857 to correct it; the revised text was published by the Admiralty as the first edition of the Persian Gulf Pilot in 1864. The Persian Gulf Pilot was a significant improvement on the reports of Colebrook and Brucks' on the geographical location of the towns of Qatar. It referred to some new places, namely, Ras Abu Aboud and Doha. While the former was a natural harbour with no settlement, the latter was a well-developed town. The Persian Gulf Pilot described the town of Doha as well as al-Bida:

> DOHEE [Doha] is a town partly walled round, with several towers, half a mile S. W. by W. from Ras Nesseh; it extends about 800 yards along the beach. The Sheikh's house is at a large round tower (with the flagstaff) on the beach, about the centre of the town; to the west of this tower is a small bight, where boats are hauled up to repair. The reef dries off a quarter of a mile from the shore opposite this place. Between this town and Al Bida', and almost connected with Doheh, is a distinct town, recently built, called Doheh el Saghireh (Little Doha), which has a new square fort at the south-west corner, built on the rising ground at the back of the town.
> AL BIDA- There is only 400 yards open space between this town and Little Doheh; the three towns together extend one mile along the coast. Bida' is built up the side of the rising ground; there is a fort in the town, where the Sheikh's flag is shown, and two towers on the highest part of the land behind the town, one of which is the first thing seen from the sea. One mile and a half to south-eastwards of the town is a tower near the wells, with a little cultivation; with this exception the whole country is desert.[12]

The total population of the above three towns was approximately 5,000; most of whom were employed in pearl fishery. The Persian Gulf Pilot also estimated the inhabitants of Wakra as 1,000.[13] Finally in 1905, John Gordon Lorimer, the eminent British civil servant and gazetteer writer estimated the total population of Qatar as 27,000 souls consisting of different tribes, namely, al-Maadhid, al Bu Ainain, al Bin Ali, al Bu Kuwara, al-Mohannedi, al-

Kubaisat, al-Dawasir, al-Manai, al-Sulaithi, the Persians, etc.[14] Lorimer also described the economic activities of Qatar at the time:

> *Occupations, resources and trade-* The principal and almost the exclusive source of livelihood in Qatar is pearl-fishing, supplemented in some places by the breeding of camels. Agriculture hardly exists. The only date palms – and they are not numerous – appear to be those in the gardens at Laqtah, Markhiyah, Mushairib, Na'aijah, Sakak, Sakhamah and Wakrah; and it is not clear that any vegetable gardens exist except at some of the same places. A few semi-wild clumps of dates are found on the west coast near Doha-as-Salwa. Besides camels the settled villagers have a few horses and cattle, which they keep in their own possession, and some sheep and goats, which are tended for them by the Bedouins. They also fish along the coast of the district.[15]

Pearl fishing, the season of which lasted from the end of May to the end of September, was the main source of income in Qatar. Pearls were exported to Bahrain, Lingah (on the Persian Coast) and Bombay in India. Lorimer stated that 817 boats were engaged in pearl fishing and about 13,000 men were employed in the pearl fishing.[16] Pearling had a very ancient history in the Arabian Gulf region and it had always been centred in the water around the peninsula of Qatar, particularly from Ras Rakkan to Khor al-Odaid. Leading tribes, namely, al-Maadhid (later al-Thani), al-Musallam, al-Bin Ali, al-Bu Kuwara, al-Bu Ainain, al-Manai, al-Mohannadi and al-Muraikhi were involved in pearl fishing and trade. The trade continued to dominate the economy in the region until the entry of the Japanese cultured pearl into the markets and the discovery of oil in the region in the 1930s. In addition to pearling, the coastal people were involved in fishing, while the traditional occupations of the settled tribes in the interior of the peninsula were herding and camel as well as horse breeding.

SHAIKHS AND TRIBES IN QATAR

The political system in Qatar was more complex than in other Gulf countries until the conclusion of the Anglo-Qatari Treaty of 1868. While there was no central administration or authority, Qatar had its own tribal system of administration, in which local tribal chiefs

7

exercised judicial and administrative powers over their own tribes. Although Brucks' report of 1829 described the tribes in Qatar, he made no reference to the tribal chiefs or a ruling Shaikh. Different towns, such as al-Huwailah, Fuwairet, al-Bida, Doha and Wakra were governed by different tribal leaders. From the mid sixteenth to the end of the eighteenth century al-Musallam tribe was predominant in the Qatar coast. However, in the absence of central government or unified leadership, from 1851 onward, Shaikh Mohammad bin Thani had been acted as de-facto chief of Qatar.[17] British sources reveal that the Political Resident Captain James Felix Jones (October 1855 to April 1861) wrote to the al 'Bida Chief' Shaikh Mohammad bin Thani asking him to restore a Lingah boat to its owner, which was done in the 'presence of Ben Sanee' (bin Thani) in August 1859.[18] Eventually, Shaikh Mohammad bin Thani emerged as the chief or ruler of the entire Qatar peninsula laying foundations of the al-Thani dynasty.[19] .

AL THANI FAMILY

Shaikh Mohammad bin Thani, was a descendent of the Maadhid tribe which was offshoot of the Beni Tamim. Towards the end of the seventeenth century, the Maadhid migrated from al-Ashayqir in Washam of central Arabia (i.e. Nejd) to Jabrin Oasis. Later, in 1710 they left the place because of the drought and proceeded to al-Sikak first, then Salwa, the base of the Qatar peninsula, from where they moved to Kuwait for a short period. Eventually around 1750, the Maadhid arrived at Furaiha (later Zubara).[20] In 1780s the Maadhid under the leadership of Thani bin Mohammad moved to Fuwairet from Furaiha because of their differences with al-Khalifa over the latter's expansionist policy at Zubara. Upon the death of Thani bin Mohammad bin Thani bin Ali, his son Mohammad bin Thani who was born at Fuwairet, became the chief of that place.[21]

Shaikh Mohammad bin Thani's position as 'the Chief' of Qatar received legal recognition when in September 1868 he signed a treaty with Colonel Lewis Pelly (November 1862 to October 1872), the Resident in the Gulf, bringing a significant change in the political status of Qatar. The treaty referred to Shaikh Mohammad bin Thani as the 'principal Chief of Katar'.[22] In fact, the 1868 Treaty reconfirmed the existence of a local leadership in Qatar similar to that elsewhere in the region. Although each tribe enjoyed autonomy in its own domain, the tribes agreed that the contacts on any matter with the Bushire Residency would be made by Shaikh Mohammad bin Thani on behalf of all the components of the tribal system of Qatar. The Agreement of 1868 also referred to all the leading tribes of Qatar: the al Mohannedi of Dhakira, al Bu Ainain of Wakra, al Bu Kuwara of Semaishma and Fuwairet, the Sudan of al-Bida, the Maadhid of Doha and al-Bida, the al Musallam of al-Huwailah and the Naim, who lived in the neighbourhood of Zubara and moved to Bahrain during the hot season, the Amamarah, who lived in Doha and Wakra, and the al Kaleb, whose location is unknown.[23] In addition to these tribes, there were also some other tribes namely; al-Manai , al-Marri, Beni Hajir and al-Manasir.

The Ottoman presence in al-Bida from 1871 to 1915 brought no tangible change in the internal power structure of Qatar. While al-Thani remained supreme in the tribal system of administration, the Ottoman garrison at al-Bida fort symbolised their nominal authority only. This was defined by the British authorities in the Gulf:

> El Bida is the chief town of El Katar [Qatar], the most southerly district on the coast of the Persian Gulf over which the Turks have a nominal authority. The status of the Chief of El Bida Jasim Ibn Thani, is doubtful, for while the Turks have appointed him a Kaimakam or deputy Governor, the Resident in the Persian Gulf continues to deal with him as an independent Chief.[24]

Shaikh Jasim Bin Mohammad bin Thani emerged as the supreme power in Qatar during the Ottoman period and acted as an independent Arab Shaikh, although his work was hampered by the presence of the Ottomans, who reported all his doings to the Ottoman authorities at al-Hasa. This was confirmed by T. Nesham, commander of *Woodlark*, who met Shaikh Jasim in January 1882.[25] In fact, a central authority had gradually emerged in Qatar in addition to the traditional tribal system of administration, in which an area or sphere of influence was controlled by a tribe or segment of a tribe.

THE MAJOR POWERS

The leading regional and international powers that shaped the history of Qatar and contributed to its emergence as a sovereign state were the Wahabis, Omanis, Utub, Ottomans and the British. The Persians and the Portuguese exercised no control over the affairs of Qatar and played no role in the modern history of the peninsula, despite the former's occasional attempts to subdue Zubara.

While the influence of the regional and international powers in the Gulf as well as in Qatar had diminished, Britain remained the predominant power in the region. Formal British relations with the Gulf region can be conveniently dated to the year 1763, when the East India Company established its first Residency at Bushire, following the conclusion of an agreement with the Persian King Karim Khan Zand. Initially, British interest in the Gulf was confined to trade and commerce and keeping other European powers from establishing a presence in the area. By the nineteenth century Britain had tremendous influence in the region and had established a predominant commercial position in the Gulf ports.[26] However, the later British presence in the Gulf was closely bound up with its status as part of an important route to Britain's Indian empire. British commercial and political interests in the region were safeguarded by the Political Residents, who were subordinate to the British government of India, and by the naval ships of the East India Company until it ceased to exist in 1863. In contrast to its rule in British India, Britain controlled the Gulf littoral states through treaties, commencing with the General Peace Treaty with the Trucial states and Bahrain in 1820 and culminating in the conclusion of the protection treaty with Qatar in 1916. The prime objective of this British treaty system of control in the Gulf, including Qatar, was to ensure peace at sea, eliminate maritime warfare and suppress the slave trade. Eventually, the British came to dominate the Gulf region, arbitrated conflicts amongst the Gulf states and acted as the guardian of the Gulf. On 3 May 1871 the British Foreign Secretary redefined British policy in the Gulf:

> It should never be forgotten that, in all its main features our position, in the Persian Gulf is one which we have taken up on grounds of policy. Its foundation in Treaty is of the most meager and narrow kind. In reality it is a position which we have arbitrarily assumed as an act of State or conquest, the justification of it lying, first, in the circumstances of the time at which we entered on it; and, secondly, the length of time during which we have asserted the position we have how exceedingly narrow is the scope of the Treaty engagements with the Arab tribes for the

maintenance of the peace at sea. The Chief of Bahrain is bound by his special agreement to abstain from all maritime aggressions of every description so long as he is protected by the British Government from similar aggressions by the Chiefs and tribes of the Gulf. But the Chief who have subscribed the perpetual peace have bound themselves only to abstain from aggressions on the subjects of the subscribing parties, and our right or duty to interfere in virtue of the Treaty is limited to cases in which one of the parties to the maritime peace commits an outrage on another. In all other cases our interference with warlike operations, whether by parties to the Treaty of peace, or by Chiefs who are not parties to that Treaty, is based on no better foundation than our own determination to do so, and the fact that we have done so for a long series of years. If we have no Treaty right, and are under no Treaty obligation to prevent such an expedition as the present on the part of Turkey or Persia, neither we have any such right, or are we under any such obligation, to prevent any of the subscribers to the maritime peace from attacking the Turks or the Persians by sea, or fitting out a similar expedition against any party who has not signed the Treaty.[27]

In other words, British policy in the Gulf was based on interpretation of the treaty system and on British will. Obviously, the peninsula of Qatar was also part of that system.

During the nineteenth century, British interests in Qatar, like its interests in the Trucial coast, were protected by the Political Residents at Bushire as well as by British war ships. With the establishment of the British Political Agency at Bahrain, the Political Agent who was subordinate to the Political Resident at Bushire became supreme in overseeing the affairs in Qatar. The following chapters prove how British involvement in the affairs of Qatar shaped the course of its history against the background of other interested parties, particularly the Utub, the Wahabis and the Ottomans.

Notes.

1.	C. G. Constable and Lieut. A. W. Stiffe, (comp.), The Persian Gulf Pilot including the Gulf of Oman, 1864 edition, vol. 1 Oxford: Archive Editions (reprint) 1989, p. 1.

2.	Peter Vine and Paula Casey, The Heritage of Qatar, London: IMMEl publishing, 1992, pp. 11-17.

3.	Ibrahim Abu Nab, Qatar: A Story of State Building, Beirut: Name of publisher is unknown, 1977, pp. 62-63.

4.	Beatrice De Cardi (ed.), Qatar Archaeological Report: Excavation 1973, Oxford: Oxford University Press, 1978, pp.184-185; see also Vine and Casey, op. cit., p. 185.

5.	Vine and Casey, op. cit., p. 185.

6.	Carsten Niebuhr's map of the Persian Gulf showing 'Guttur' [al-Bida], al-Khor, al-Huwailah, Yusufiyah and Furaiha of Qatar in 1765, see Map no. 2.

7.	R. M. Webster, 'The Nature and Status of Qatar at the time of the Anglo-Ottoman Convention of 1913, an unpublished Research paper, Exeter (U. K.), University of Exeter, 1990, p. 10.

8.	As quoted in J. A. Saldanha, The Persian Gulf Précis: Précis of Turkish Expansion on the Arab Littoral of the Persian Gulf and Hasa and Katif Affairs 1804-1904, vol. V, Buckinghamshire (U.K.): Archive Editions, 1986, pp.2-3.

9.	G. B. Brucks, Memoir Descriptive of the Navigation of the Gulf of Persia, in R. Hughes Thomas (ed.), Selections from the Records of the Bombay Government, No. XXIV, (hereafter cited as Bombay Selections), New York: Oleander Press, 1985, p. 559.

10.	Ibid., p. 559.

11.	For more details see ibid., pp. 560-564.

12.   Constable and A. W. Stiffe, op. cit., pp. 104-105.

13.   Ibid., p. 105.

14.   For details on the Tribes and the Towns of Qatar, see Lorimer, op. cit., Geographical and Statistical, vol. IIB, pp. 1530-1532.

15.   Ibid., p. 1532.

16.   Ibid., p. 1533.

17.   Lorimer, op. cit., Historical, vol. IB, p. 800.

18.   For more details see 'Affairs claiming Resident's attention during the Tour of the Arabian Gulf Coast', unnumbered, (exact date is not known) 1860, R/15/1/166.

19.   William Gifford Palgrave, Narrative of a Year's Journey Through Central and Eastern Arabia (1862-63), vol. 2, London: Macmillan, 1865, p. 232.

20.   See the Geographical Table of the al-Thani (Maa'Dhid) family of Qatar in Lorimer, op. cit., Historical vol. 3, Tables-Maps, Pocket no. 8; see also Prideaux to the Resident in the Persian Gulf, no. 455, 23 December 1905, R/15/2/26. For the actual year of bin Thani's arrival at Qatar, see Haji Abdullah Williamson's 'Political Note on Qatar' in Colonel H. R. P. Dickson, Political Agent, Kuwait to Col. Trenchard Craven Fowle, Political Resident in the Persian Gulf, no. C-17, 18 January 1934, F. O. 371/17799.

21.   Mustafa Murad al-Dabbagh, Qatar: Madiha wa Hadiruha, Beirut: Dar al-Taliha, 1961, p. 176.

22.   For full text see C. U. Aitchison, A Collection of Treaties, Engagements and Sanads, vol. XI, Delhi (India): Manager of Publishing, 1933, p. 255.

23.   Lorimer, op. cit., Geographical and Statistical, vol. IIB, pp. 1530-1531.

24. P. I. C. Robertson, Assistant Political Agent, Basra to Political Resident, Baghdad, no. 255, 2 September 1886, R/15/1/187.

25. T. Nesham, Commander HMS 'Woodlark' to E. C. Ross, Resident in the Gulf, unnumbered, 7 February 1882, R/15/1/187.

26. For more on British Gulf Administrative System, see Penelope Tuson, The Records of the British Residency and Agencies in the Persian Gulf, London: India Office Library and Records, 1979, pp. 1-9.

27. As quoted in J. A. Saldanha, op. cit., Précis of Turkish Expansion on the Arab Littoral, of the Persian Gulf and Hasa and Katif Affairs, vol. 5, p. 15.

CHAPTER TWO

The Shifting Balance from Zubara to Al-Bida

Zubara's origin as a seaport is not known although during the eighteenth century, it emerged as a centre of pearl trade to India and an important transit point for trade and commerce between East and West. The economic prosperity and the prospects of this port town continued to grow and in the 1760s it attracted foreign merchants such as the Utub, Rizq al Asad and Jalahima branches of the Utub tribe, who migrated to this land of fortune from Kuwait. The foreigners soon monopolised not only trade and commerce but also political power in the town, driving out the original settlers. Nevertheless, all the leading Qatari tribes, including the displaced leaders of Zubara, buried their differences and joined with the Utub in the conquest of Bahrain. In the 1890s Zubara began to decline and became a power base not only for the Utub but also for the Wahabis and the dislodged Omanis. Zubara was destroyed by the Sultan of Oman (Muscat) ultimately. The fall of Zubara together with Khor Hassan, the head quarters of Rahma bin Jaber, led to the emergence of al-Bida as a focal point for commerce and politics.

THE FOUNDATION OF ZUBARA
The Rise and fall of Zubara should be narrated against the background of the advent of the Portuguese power in the Arabian Gulf and Portuguese control of the Strait of Hormuz from 1515. Portuguese ascendancy in the Gulf marked a turning point in the emergence of Qatar in general and Zubara in particular. Ancient or mediaeval civilizations and settlements grew on the banks of rivers

or sea-coasts and the decline or destruction of one nurtured the growth of another. The same pattern might be applied to the vital western coast of Qatar. A new civilizaton or, more modestly, a settlement grew and developed along the shore of the western coast from the ashes of the destruction of coastal villages. A Portuguese naval squadron of six vessels under the command of Captain D. Goncalo da Silveira set 'fire to villages along the shores of Qatar', apparently on its way to invade the Persian King's possession of Bahrain between September 1627 and April 1628.[1] Neither Silveira nor any other authority specified the location of these torched and destroyed coastal 'villages of Qatar'. They may well have been places like al-Bida, al-Huwailah, al-Khor and Fuwairet given their location and substantial population. It is possible that the survivors moved to Yusufiyah and Furaiha or Zubara and built settlements there because of their strategic locations and proximity to the sea.

The growth of Zubara thus apparently resulted from the dislodging of the inhabitants from the other coastal settlements. The settlers, particularly the al-Musallam, a leading Arab tribe originating from the al- Tamim tribe, to which al-Mundhir ibn Sawa al-Tamimi, the first known ruler of Qatar in 628 AD, also belonged, might have moved to Furaiha from al-Huwailah. It is worth noting that the town of Furaiha, which was situated about two miles to the north-east of Zubara, was more prominent than that of the latter during that time. However, gradually Zubara began to grow under the leadership of al-Musallam.[2] Although no genealogical records are extant and the names of the leaders of this powerful tribe are unknown, it is evident from an Ottoman source that in 1555, one Mohammad bin Sultan beni Muslim (al-Musallam) was the Shaikh of Qatar with his headquarters at al-Huwailah. This Shaikh also owned land at al-Hasa of Eastern Arabia with which he maintained close links. This is the oldest Ottoman record relating to the origin of Qatar.[3] However, the al-Musallam later became tributary to the Beni Khaled (a branch of Beni Tamim), who gained pre-dominance in Eastern Arabia and established their headquarters at al-Hasa in the first half of the eighteenth century, under the leadership of Salman bin Mohammad. Having established their influence on the Qatar coast through the al-Musallam, the Beni Khaled extended their authority westwards, up as far as Kuwait, by the middle of the eighteenth century.[4]

Gradually, Zubara developed as a commercial port under the leadership of the al-Musallam and soon drew the attention of other tribes such as the Utubs from the Beni Khaled's domain of Kuwait. In the year 1766, the al-Khalifa branch of the Utub, having failed to expand their business in Kuwait and quarrelled with their kinsmen, the al-Sabah, moved to Basra and later arrived in Zubara under the leadership of Shaikh Mohammad bin Khalifa. He was accompanied by Rizq al-Asad, a leading Basra merchant. Zubara's geographical position and its emergence as a sea port made it an attractive prospect for these migrants. Zubara was not unknown to the al-Khalifa; they first came in 1710 along with al-Sabah branch of the Utub, when the town was under the control of the al-Musallam. However, their stay at Zubara during that time was brief and they left for Kuwait in search of better opportunity.[5]

The al-Musallam raised no opposition when the al-Khalifa landed at Zubara for the second time. However, their relations gradually soured owing to al-Khalifa's ambition and expansionism. Inter tribal tension peaked when the al-Khalifa declined to pay tax to the al-Musallam and attempted to strengthen their position by expanding their settlement, constructing a wall and fort called Qalat Murair, one and half miles south-east of Zubara port. The construction of this fort was completed in 1768, about two years after their arrival at Zubara. Qalat Murair was connected 'with the sea by a creek, which enabled sailing boats to discharge their cargoes at its gate'. In addition to the Murair fort, Zubara already featured a number of towers and well-defended forts built by earlier settlers.[6] It should be noted that in addition to the al-Musallam, two more prominent tribes, the Maadhid and al-Abu Hussain had also been living in the neighbouring Furaiha and Yusufiyah respectively even before the arrival of the al-Khalifa. While al-Abu Hussain was another branch of al-Musallam, the Maadhid was an offshoot of the Beni Tamim like the al-Musallam[7] and the Maadhid also came to Qatar coast from Kuwait in about 1750. Like that of the Utub they were also originally from central Nejd from the Jabrin Oasis. This was reconfirmed by Shaikh Jasim bin Thani, the second ruler of Qatar, later in 1905:

> Shaikh Jasim-bin-Thani gives the pedigree of his grandfather as follows:- Thani bin Muhammad bin Thani bin Ali bin Muhammad bin Salim bin Muhamamd bin Jasim bin Sa'id bin Ali bin Thamir bin Muhamamd bin Ali bin Ma'dhad bin Musharraf. He says that Ma'dhad governed the Jabrin Oasis, whence the family, at a later date,

removed on account of the unhealthiness of the climate; after a sojourn at Kuwait they finally settled in Qatar.[8]

THE RISE OF ZUBARA

Zubara, when the al-Khalifa arrived, was a well developed town in which the three tribes mentioned above, occupied the pre-eminent position. Within less than a decade of their arrival, the al-Khalifa came to dominate the town, forcing the al-Musallam to retire to al-Huwailah and the Maadhid to Fuwairet, where they joined the al-Bu Kuwara, their kinsmen and a branch of the al-Bin Ali. The clash between the al-Khalifa immigrants and the leading local tribes in Zubara and Furaiha took place because of commercial interests of the al-Khalifa. The al-Khalifa's primary objective was to gain full control of Zubara and its port to serve their commercial interests; the town had developed strong trade links not only with Muscat and the Indian ports of Surat and Calcutta in the East but also with Kuwait and Basra in the West. Furthermore, Zubara had also maintained special commercial relations with the port of Qatif.[9]

Although, no statistics are available for monthly or annual trade in Zubara, the following mercantile goods were imported through the port:

> ...Surat Blue and other Piece Goods, Cambay Chauders, Guzerat Piece Goods and Chintz, Shauls, Bamboos and many less important mercantile Articles from Surat, etc. Spices, Bengal Soosies, Iron, Lead, Tin, Oil, Ghee, Rice and many less important mercantile Articles from Muscat; and of Dates and of Grain from Bussora [Basra].[10]

Most of Zubara's imports, for consumption locally or in the immediate vicinity, started their journey from the Indian medieval ports of Surat, Calcutta and Chittagong in Bengal. Some imported items were re-exported as far as Kuwait and Basra, from where local merchants again re-exported certain goods in caravans to Baghdad and Aleppo (Syria).[11]

Having monopolised the trade of Zubara, the al-Khalifa abolished taxes and turned the Zubara port into a free trade zone. This policy severely affected the two neighbouring ports at Oqair and al-Hasa, as well as distant ports such as Muscat and Basra, which were collecting import duties on all mercantile importations.[12] Consequently, Zubara became a favourite transit point for merchants transporting goods from India to Syria and other Ottoman territories.

Zubara's economic growth encouraged the Jalahima section of the Utub tribe to migrate there from Kuwait under the leadership of Jaber bin Utub. The al-Khalifa, however, refused to share their economic gains with their kinsman Jaber; their hostile attitude towards the Jalahima eventually forced the latter to abandon Zubara and settle at Ruwais.[13] In fact, Zubara fell victim of Jalahima-al-Khalifa economic rivalry and eventually the Jalahima chief was killed in an armed clash with the al-Khalifa, after which the influence of the al-Khalifa increased rapidly in and around Zubara, including Khor Hassan and Ruwais. The al-Khalifa gradually monopolised the pearl banks off the coast of Qatar. Zubara's economic prosperity received a further boost during the period from 1775 to 1776, when the Persians occupied Basra, diverting much of its trade as well as that of Kuwait and forcing several leading Kuwaiti and Basra merchants including Ahmed Sadoun of Kuwait to move to Zubara.[14]

ZUBARA'S OCCUAPTION OF BAHRAIN

The emergence of Zubara as a transit port and its increased commerce became a cause of anxiety for the Arabs of the Persian coast, who were determined to destroy the port. It was Ali Murad Khan, the ruler of Shiraz, who masterminded a plan empowering Shaikh Nasir bin Madhkhor of Bushire, who was also the governor of Bahrain, to destroy Zubara. Shaikh Nasir, aided by the Beni Kaab and the Arabs of Bandar Riq, made a series of raids on Zubara between 1777 and 1801. However, the attacks were repulsed by the combined forces of Zubara and Arabs from elsewhere on the Qatar coast.[15] To frustrate Shaikh Nasir's designs on Zubara, the people of the town retaliated with a surprise attack on the island of Bahrain in September 1782. After a brief encounter the raiders forced Shaikh Nasir to retreat to the fort. After plundering and destroying the town of Manama they returned to Zubara with a Jalibut belonging to Bushire.[16]

Shaikh Nasir was not the person to give up easily. After the defeat, he seething with anger, plotted his next move against Zubara with the aim of destroying it. Ali Murad Khan and the rulers of other Persian coastal provinces, like Bunder Righ, Ganaveh and Dahistan also joined with Shaikh Nasir. An expeditionary force consisting of 2000 men under the command of

Mohammad, a nephew of Shaikh Nasir, proceeded to the Qatar coast to invade Zubara in December 1782. Blockading the port of Zubara for about five months, the Persian forces landed on 17 May 1783 to storm the Murair fort. Although they expected to accomplish their task with little or no resistance, to their great surprise no sooner had they moved forward than a large number of Zubaran troops emerged from the fort and attacked them. Fierce fighting ensued and soldiers were killed on both sides. The Persians were in disarray when Mohammad, the commander of their forces, was killed. Eventually, they threw down their arms, fled to their ships and sailed for Bushire.[17] The Zubaran forces could not pursue the Persians as they lacked sufficient naval power.

Upon hearing of the distress of their kinsmen at Zubara, the al-Sabah of Kuwait despatched a fleet of six Jalibuts and a number of armed boats to assist them. The Kuwaiti contingent landed on the coast of Bahrain and obliged the force, left by Shaikh Nasir to protect the island, to withdraw to the fort.[18] As soon as news of the successful Kuwaiti operation reached Zubara, the people of the town made every effort to use the opportunity to occupy Bahrain. They raised a strong force consisting of Jalahima from Ruwais and various tribes of Qatar, which included:

> Al Musallam from Huwailah, Al Bin-Ali from Fuwairat, Sudan from Dohah, Al Bu Ainain from Wakrah, Kibisha [Kubaisat] from Khor Hasan, Sulutah [Sulaithi] from Dohah, Mananaah [Manai] from Abu Dhuluf, Sadah from Ruwais, Al Bu Kuwarah from Sumaismah and Naim Bedouins from the interior of the promontory.[19]

Following their landing at Bahrain, the combined forces joined with the Kuwaitis and besieged the Persian garrison at Manama. After a siege of about two months the Persians capitulated on 28 July 1783, and were permitted to go to Bushire. Thereafter, Bahrain became the seat of power of al-Khalifa under the leadership of Ahmed bin Khalifa, although some members of the family continued to frequent Zubara.[20] Despite the continuous fighting and instability, Zubara's commerce flourished and the port grew to be larger and more important than Qatif by 1790. Foreign traders still enjoyed the privilege of tax exemption.[21]

The victorious forces soon divided into two opposing groups. Although, all the major tribes of Qatar had played a vital role in the war with Bahrain, the al-Khalifa tribe, which numbered only four persons,[22] capitalised on the fortunes of victory. As the al-Khalifa refused to share the territorial and political gains with other tribes, the Jalahima, who also actively participated in the conquest of Bahrain, took up arms against them. Consequently, the Jalahima, along with other Qatari tribes except the al-Bu Kuwara, al-Sulaithi and al-Musallam, returned to Qatar in disgust. The Jalahima family, under the leadership of Rahma bin Jaber, the son of Jaber, who led the original migration from Kuwait to Zubara, established their headquarters at Khor Hassan and organized a strong force to fight against the al-Khalifa.[23]

WAHABI RULE AND THE FALL OF ZUBARA
Meanwhile, a new movement had emerged in Arabia which changed the course of history for the whole region, including Qatar. This was the Wahabi movement founded by Abdul Wahab and his son Mohammad bin Abdul Wahab in 1742. The founders of this sect had established good relations with the Shaikhs of the Saudi family who then ruled at al-Diriyya in Nejd. Although the movement was originally religious in character and advocated a return to the 'positive purity' of Islam, gradually it acquired a political complexion under the patronage of Mohammad bin Saud, ruler of al-Diriyya, and the movement grew rapidly in Nejd. In fact, bin Saud was the first Wahabi Amir. On his death in 1765, he was succeeded by his son Abdul Aziz, who extended Wahabi influence as far as Riyadh in 1772, making it tributary to him. Having subdued Riyadh, Abdul Aziz, under the command of Ibrahim bin Ufaysan, conquered al-Hasa in 1795, diminishing the Beni Khaled's power.[24]

The decline of Zubara in the 1790s should thus be seen against the background of the fall of al-Hasa. It is worth noting that Zubara was a fortified town not very far from al-Hasa. Therefore, many leading inhabitants of al-Hasa, including Barrak Abd al-Muhsin and Duwayhis, the two Beni Khaled leaders, took refuge there because of the historical links between the two places.[25] The protection afforded to the dislodged Beni Khaled by the people of Zubara worried the Wahabis, who thought that the Beni Khaled and the Zubarans might conspire against the Wahabi regime at al-Hasa. With this in mind, the Wahabis were determined to bring Zubara under their control. Nevertheless, the refugees were not the main reason why the Wahabis turned their eyes towards Zubara

21

and began making military plans. The Wahabis thought that the Zubara Utub were engaged in *shirk* (pluralism) and *bida* (innovation), which went against their teaching. Zubara had also grown prosperous over the years. This together with its strategic location as gateway to the East and West, added fuel to Wahabi expansionism. With these objectives in mind the Wahabis, who lacked naval power, raided Zubara in 1793 and applied guerrilla tactics.[26] Eventually, by 1795, under the command of Ibrahim bin Ufaysan, the Wahabis conquered Zubara, forcing the inhabitants, including Shaikh Salman bin Ahmed al-Khalifa, who was in Zubara at the time, to flee to Bahrain. Having gained full control of Zubara, bin Ufaysan established his ascendancy over the neighbouring towns, namely Furaiha, al-Huwailah, Yusufiyah and al-Ruwayda.[27] The fall of Zubara was apparently due to its failure to enlist any support from other Qatari tribes, who were annoyed at the behaviour of the al-Khalifa. Outside military assistance from Bahrain island or their kinsmen in Kuwait failed to materialise, adding to the Zubarans' misery.[28]

There was no immediate end to the al-Khalifa's misfortunes. In the year 1799, Sayyid Sultan bin Ahmed, the Sultan of Muscat, attacked Bahrain, claiming that boats owned by the al-Khalifa were avoiding paying tax for passing through the Strait of Hormuz. Having failed to occupy the island, Sayyid Sultan returned to Muscat and launched another vigorous attack in 1800. This time he easily overran Bahrain, arresting all the headmen and sending twenty-six leading members of al-Khalifa families to Muscat. However, some al-Khalifas escaped to Zubara and sought Wahabi protection, which was readily granted. In 1801, 'assisted by all the Wahabee dependants in the district of Kutter' (the Qatari tribes) the exiled al-Khalifas attacked and recaptured Bahrain, having forced the Sultan's governor and his son to flee the island.[29] Once again, therefore, the Qartaris extended their assistance to the al-Khalifas, the occupation of the island of Bahrain was achieved as a result of their active support.

Having established their influence in Bahrain, by 1802 the Wahabis had completed the task of reducing to nominal submission the whole coast from the mouth of the Shat al-Arab to the northern limits of Muscat territory, including the Qatar coast. An interesting 'chronological précis of Katr History', prepared by Captain Francis Beville Prideaux, Political Agent in Bahrain (October 1904 to May 1909) in 1905, more than one hundred years after the Wahabi occupation of Zubara, revealed that in 1802, Abdullah bin Ufaysan was entrusted with the collection of revenue from Bahrain, Zubara, Khor Hassan, and Qatif, and Wahabi garrisons were maintained in all these places, though the local chieftains were permitted to continue carrying out administrative tasks. The report added that the Jalahima Shaikh of Khor Hassan, Rahma bin Jaber, who had most ingratiated himself with the Wahabis, was permitted to assume possession of Zubara. The report was produced after extensive research in connection with the preparation of Lorimer's *Persian Gulf Gazetteer*.[30] However, the Wahabis continued to dominate Zubara despite the granting of local autonomy and bin Jaber did not take over the charge of Zubara.

The Wahabis' involvement in dynastic rivalry in Muscat brought disaster for Zubara. In 1803, Sayyid Badar, son of Sayyid Sultan, failed to seize the fort of Muscat during the Sultan's absence on a pilgrimage to Makkah and fled to Zubara, where he obtained the protection of the Wahabi Amir. Sayyid Badar raised a strong force and captured Muscat in 1805, with the assistance of Zubara Arabs as well as strong Wahabi support. Sayyid Badar ruled Muscat until he was killed by his cousin Sayyid Saeed in 1807. Consequently, Wahabi influence in Muscat declined,[31] prompting the people of Zubara to seek an alliance with Sayyid Saeed. They even endeavoured to agree a definitive treaty with the new regime in Muscat in 1809, as Bombay official, Francis Warden, noted:

> The Uttoobees at Zobara had also suffered so much from the tyranny of the Wahabees that they had recently endeavoured to conciliate the Government of Muskat, frequented the port, and paid duties as other States; whilst the Imaum, with a judicious policy, showed a preference to them, in remitting such duties as might fall heavily on their trade. No obstacle, therefore, opposed the conclusion of a solid agreement between the Uttoobees [Zubarans] and the Government of Muskat, but the want of a guarantee [sic], who could secure the due performance of its stipulations.[32]

In other words, no defensive alliance was made between Zubara and Muscat because no means could be found to guarantee its effectiveness and implementation.

Zubara's lack of support for the Wahabis and the refusal of the al-Khalifa of Bahrain to join the Wahabi-Qawasimi coalition to wage war at sea and launch expedition against Basra, angered the Wahabi Amir Saud. Late in 1809, Saud summoned the leading members of the al-Khalifa family, who had intrigued against the Wahabis and who maintained an influence in Zubara to the Wahabi capital from Bahrain. Others members of the family, however, escaped to Muscat to apply for military assistance from Sayyid Saeed. The Wahabi chief's action was prompt and decisive. Taking control of all of Zubara and Bahrain, he appointed Abdallah bin Ufaysan and Fahad bin Sulaiman bin Ufaisan respectively as governors to supervise and administer the affairs of Zubara and Bahrain on his behalf.[33]

Wahabi administration of Zubara and Bahrain was short lived. In early 1811, taking advantage of the Wahabi Amir's withdrawal of his garrison from Zubara for deployment against the Egyptian commander, Ibrahim Pasha, who was advancing towards Jeddah and Makkah, Sayyid Saeed sent an expedition against Zubara with the collaboration of al-Khalifa exiles and under the command of his brother, Saeed Waliullah bin Hamid. The invaders torched and levelled the town.[34] Having accomplished their mission in Zubara, the Muscat troops moved to Bahrain and captured the Wahabi vessel, *Duryah Begee*, with fifteen of their principal officers. The Wahabi governor was also arrested and the al-Khalifas restored to power, but only as subordinates to Muscat.[35] From Bahrain, the Muscat force once again headed for the Qatar coast and attacked Khor Hassan, the headquarters of bin Jaber. Consequently, bin Jaber moved to Dammam on the coast of al-Hasa, with the agreement of the Wahabis.[36]

THE ASCENDANCY OF BIN JABER

Although little is known about the history of the Qatar coast from 1811 to 1820, the period was dominated by bin Jaber's ascendancy throughout the Gulf including Qatar. The history of Qatar would thus be incomplete without mention of the story of bin Jaber's adventures in the Gulf. The war of attrition against the al-Khalifa of Bahrain began when the Zubarans attacked Bahrain, occupied the island and the al-Khalifa squeezed out the Jalahima section of the Utub as mentioned above. Bin Jaber continued his campaigns against the trade and shipping of Bahrain with the backing of the Wahabis and the people of Muscat. His base at Khor Hassan became a sanctuary and gathering place for all anti al-Khalifa forces. The geographical position of Khor Hassan served as a natural defence against any outside attack on bin Jaber:

> ...Rahmah's base at Khaur Hassan [now known as Khuwair] was a mean and dilapidated village composed mainly of *barastis,* or primitive huts. A square fort, built of coral and mud, commanded the beach and the shallow anchorage. Off-shore there were two coral reefs, visible at low water, which ran north and south, parallel to the coast, with a channel of 1½-2 fathoms in depth and about 2 cables' length in breadth between them, down which vessels could sail from Ras Rakan and on to Zubara. A channel through the reefs about two miles west of the harbour was navigable only at spring tide, and could easily be blocked with stones when attack threatened. Rahmah had a second lair to which he could retire, further down the coast at Dauhat al-Husain.[37]

In addition to this, vessels used to anchor at Khor Hassan from 4 to 6 fathoms water, and shelter from all winds but those that blow from the north.[38] Because of those advantageous conditions in and around his base, bin Jaber's operations against his arch enemy the al-Khalifa were successful.

While bin Jaber adopted a non compromising attitude towards the al-Khalifa, he was careful not to antagonize the British and avoided any encounter with the British vessels in the Gulf. Bin Jaber's pragmatic policy towards the British kept the latter from entering into confrontation with him. Lorimer commented:

> The exploits of Rahmah, though in some cases piratical, were performed as a rule under pretext of lawful warfare; and towards the subjects and officials of the British Government, even at a period when no respect was shown for them by the Qawasim, his conduct was scrupulously correct. On one occasion he was stated to have shown 'remarkable ... forbearance towards the Augusta Cruizer when in the power of his fleet'.[39]

However, the British attitude towards bin Jaber changed significantly following their expedition in 1809 against the Qawasim, who were involved in attacks on British ships passing through their territorial waters. When several Qawasimi vessels took shelter at Khor Hassan, escaping destruction at Ras al-Khaimah, bin Jaber extended all kinds of assistance to them.[40] Bin Jaber's pro-Qawasim leanings angered the Bombay government. Therefore, in early January 1810, Captain Wainwright and Colonel Lionel Smith were instructed to obtain an undertaking from bin Jaber that he would desist from assisting the Qawasim. If he refused to do so, Khor Hassan and its harbour would be bombed.[41] General John Malcolm, the leader of a mission to Persia brought the instruction but at a meeting on HMS *Psyche* on 30 January 1810, agreed with Wainwright and Smith that an attack on Khor Hassan would be a difficult venture, 'full of danger, if not impracticable' because the winter north-wind was blowing and large war ships were unable to approach the harbour.[42]

Nicholas Hankey Smith, the brother of Colonel Lionel Smith and the Political Agent at Muscat (January 1810 to April 1810), who also attended the meeting, held a significantly different view from Malcolm and the two expedition commanders. He was convinced that although bin Jaber had previously avoided attacking British vessels he would change his policy when the British 'armament quitted the Gulf'. He had two reasons for his contention:

> 1st. The natural ferocity of Rahma's Character, in proof of which some instances of his cruelties and depredations are adduced.
> 2ndly. Rahma's great accession of Power in consequence of the Wahabee having ceded, to him Bahrein and Zoobarah; besides, his having been joined by the Joasmee fugitives, which had increased his fleet to 40 Vessels and Boats.[43]

Smith firmly believed that bin Jaber, with his increased power, would soon emerge as the 'terror of the Gulf'. Smith argued that Khor Hassan, along with bin Jaber's entire fleet, should be destroyed by the expedition. He believed this would force bin Jaber to surrender.[44]

Smith's arguments for destroying Khor Hassan failed to convince governor Jonathan Duncan, who, endorsing the decision reached by Malcolm and the two other military personnel, maintained that any attack on bin Jaber's headquarters at Khor Hassan would 'vitiate the principle of reprisal upon which the operations in the Gulf had been based'.[45] In view of Duncan's opposition and on Malcolm's advise, Smith dropped the idea of destroying Khor Hassan. He now sent a letter of 'friendly admonition' to the Wahabi chief Saud bin Abdul Aziz, requesting him to restrain bin Jaber from giving shelter to the Ras al-Khaimah fugitives. The letter was carried to Khor Hassan in March 1810, by Captain N. Warren and accompanied by Lieutenants Eatwell and Frederick of HM 65th Regiment, in the cruisers *Vestal* and *Ariel*. The letter was probably delivered to Abdullah bin Ufaisan, the Wahabi Agent on the Qatar coast, who was then residing just a few miles away from Khor Hassan in Zubara. During their visit to Khor Hassan the two naval officers observed that bin Jaber's stronghold and Khor Hassan's natural defensive advantages and inaccessibility would make an attack very difficult.[46]

THE FLIGHT OF BIN JABER

It is an irony of history that Smith's dream of destroying bin Jaber's stronghold was realised not through British action, but by Muscat's attack on Khor Hassan. This even compelled bin Jaber to settle at Dammam on the coast of al-Hasa, from where he continued to attack al-Khalifa ships. Bin Jaber, who had been representing the Wahabis at sea, was greatly affected by the sudden demise of the Wahabi Amir Saud bin Abdul Aziz in 1814, and the subsequent decline of Wahabi power. The continued success of the Egyptian expedition and the absence of forceful and strong leadership after the death of Saud bin Abdul Aziz was a further blow to Wahabism. Bin Jaber's relations with the new Wahabi Amir, Abdullah bin Saud, became strained as a result of the al-Khalifa's alliance with the new Wahabi leadership, intended to reduce Muscat's domination in Bahrain. In 1816, having deserted the Wahabi cause, bin Jaber joined with the Sultan of Muscat and made an unsuccessful attack on Bahrain. Bin Jaber's change of loyalty, however, infuriated the Wahabi Amir: in July 1816, bin Jaber's fort at Dammam was blown up by the Wahabis without much difficulty. He managed to rescue his family however, and after staying briefly at Khor Hassan, in October 1816, he, along with the 500 members of his family, arrived at Bushire as fugitives. Bin Jaber brought with him a number of small boats, two large Bughlahs, and a very big Battil.[47]

27

Bin Jaber's life in Bushire differed from that he had experienced in Khor Hassan and Dammam. He was given a warm reception and all kinds of support including the allotment of a particular quarter of the town to reside in by Shaikh Mohammad, governor of Bushire. Upon his arrival, the former chief of Khor Hassan made sure to pay a visit to William Bruce, the Political Resident at Bushire (September 1808 to July 1822) and renew his declared friendship for the British government. He was ready to fight against the Qawasim if the British government wished. True to his word, in March 1817, bin Jaber successfully attacked twelve Qawasimi boats which were carrying supplies from Bahrain to Ras al-Khaimah; four of these he carried to Bushire and the remaining eight he destroyed.[48] In June 1817, bin Jaber seized two more Qawasimi vessels which, along with twenty-two other boats, had attempted to attack the East India Company's vessels, *Vestal*, *Alexander*, and *Petric*. He did this on his way back to Bushire from Muscat, where he had gone to enlist Sayyid Saeed bin Sultan's support for a new attack on Bahrain. His mission to Muscat was however, unsuccessful due to Saeed's preoccupation with various internal matters.[49]

Meanwhile, significant changes were taking place in Arabian politics. Ibrahim Pasha, son of Muhammad Ali Pasha of Egypt, occupied the Wahabi capital al-Diriyya in 1818, further diminishing Wahabi power. Bin Jaber seized the opportunity presented by the decline of Wahabi power and quickly joined with Ibrahim Pasha to regain his lost position in Dammam and attack the port of Qatif. He returned to Dammam and rebuilt his fort, which had been destroyed by the joint forces of the Wahabi and al-Khalifa in 1816. He now maintained close connections with the province of al-Hasa rather than with the coast of Qatar. While he continued his operations against Bahrain from Dammam, he maintained friendly relations with the British authorities in the Gulf. it is worth noting that in July 1819, when Captain George Forster Sadlier landed at Qatif on his mission to Ibrahim Pasha, bin Jaber rendered him 'every assistance piloting the Vestal cruiser himself into the harbour'.[50] Maintaining his customary friendly relations with the British authorities, bin Jaber continued his struggle against Bahrain. In February 1820, he joined with the Prince Governor of Fars, who was planning to invade Bahrain. But on the way one of his largest Baghlas was wrecked upon a shoal near Bardistan (on the Persian coast), forcing him to abandon his mission and return to Dammam.[51]

Failing to deter bin Jaber, the al-Khalifa adopted a policy of appeasement and started paying an annual tribute of 4000 German Crowns to him in April 1820.[52] This tribute was significant, as it means that Bahrain acknowledged the supremacy of bin Jaber, the former chief of Khor Hassan. However, as bin Jaber's policy towards the al-Khalifa remained the same despite their payment of tribute, they concluded a peace treaty with him in February 1824, following mediation by Colonel Ephraim Gerrish Stannus, the Resident at Bushire (December 1823 to January 1827).[53] By this agreement both parties 'undertook to be at peace with one another for the future', upon certain conditions including that bin Jaber agree to withdraw his protection from the Abu Sumait tribe, which had previously fled from Bahrain and taken shelter at Dammam. Nevertheless, bin Jaber still restricted Bahraini movements at sea and eventually died in 1826, in an encounter off the coast of Dammam with Ahmed bin Sulaiman, the nephew of the ruling Shaikh of Bahrain. Later, Samuel Hennell, the Acting Resident in the Gulf (June 1834 to July 1835), painted a vivid picture of his dramatic death:

> Having, therefore, given orders for his vessel to grapple with the enemy, he took his youngest son (a fine boy about eight years old) in his arms, and seizing a lighted match, directed his attendants to lead him down to the magazine. Although acquainted with the determined character of their chief, and of course aware of the inevitable destruction which awaited them, his commands were instantly obeyed, and in a few seconds the sea was covered with the scattered timbers of the exploded vessel, and the miserable remains of Rahmah bin Jaubir and his devoted followers. The expulsion set fire to the enemy's Buggalow, which soon afterwards blue up, but not before her commander and crew had been rescued from their impending fate by the other boats of the fleet. Thus ended Rahmah bin Jaubir, for so many years the scourge and terror of this part of the world, and whose death was felt as a blessing in every part of the Gulf. Equally ferocious and determined in all situations, the closing scene of his existence

displayed the same stern and indomitable spirit which had characterized him all his life.[54]

Bin Jaber, therefore, can be regarded, from one point of view, as a brave and courageous maritime leader, who sought throughout his life not only to keep the al-Khalifa away from the coast of Qatar but also to restrict their movements in the waters of the Gulf.

THE EMERGENCE OF AL-BIDA
The decline of Khor Hassan, the second major town of the Qatar peninsula during that time and the end of bin Jaber's ascendancy in the Gulf not only paved the way for the emergence of al-Bida but also opened the door to Bahrain's interference in the affairs of Qatar. While the western coast, particularly Zubara and Khor Hassan, had previously been the centres of political and commercial activity, in early 1820 al-Bida, situated on the eastern coast of Qatar peninsula, was the only port from which trading vessels regularly sailed.

Al-Bida now became the focal point. This was apparent following the conclusion in January 1820, of the General Treaty of Peace between the British East India Company and all the Maritime chiefs of the Trucial states, namely, Abu Dhabi, Dubai, Sharjah, Ajman, Umm al-Qawain and Ras al-Khaimah. The Treaty outlawed warfare by land and sea and restricted the kidnapping of slaves.[55] In the following month, Bahrain also became a signatory to the Treaty and agreed not to permit any sale of plundered or piratical goods in Bahrain.[56] Qatar was not included in the Treaty as bin Jaber, who was supposed to represent Qatar, declined to sign, arguing that he had become a protégé of the Persian government. However, the General Treaty of Peace also laid down the individual conditions for peace between Britain and each of the coastal Shaikhdoms. It also prescribed a flag for use by all friendly Arabs included in the Treaty:

> The friendly (literally the pacificated) Arabs shall carry by land and sea a red flag, with or without letters in it, at their option, and this shall be in a border of white, the breadth of the white in the border being equal to the breadth of the red, as represented in the margin (the whole forming the flag known in the British Navy by the title of white pierced red), this shall be the flag of the friendly Arabs, and they shall use it and no other.[57]

The terms of the 1820 Treaty were not applicable to Qatar as it was not a signatory. The British authorities in the Gulf therefore did not ask Qatar to fly the prescribed Trucial flag, under Article 3 of the General Treaty of Peace, as they did the lower Trucial states.

Al-Bida was in the limelight for the first time in the history of Qatar when in 1821 the East India Company's Brig *Vestal* destroyed the town on account of a maritime warfare off the coast of al-Bida. The bombardment forced three or four hundred of the inhabitants to quit the town and take shelter temporarily on the islands between Qatar and the Trucial coast.[58] In January 1823, almost two years after the *Vestal* had bombarded the town in the course of an exploratory voyage along the Arabian coast, Captain John MacLeod, the Political Resident in the Gulf (December 1822 to September 1823), paid a visit to al-Bida. This was the first time a high ranking British official had visited al-Bida and this visit provided first hand knowledge of the eastern coast of the peninsula. MacLeod met Shaikh Buhur bin Jubrun, the chief of the al-Bu Ainain tribe and observed that al-Bida was the only port, featuring a substantial number of trading vessels at the time. To his great surprise, the Resident noticed that neither bin Jubrun nor any-body else in al-Bida was aware of the General Treaty of Peace of 1820 and its terms. Although bin Jubrun and his followers expressed their willingness to abide by the terms of the Treaty and fly the Trucial flag on their boats, MacLeod did not press them to do so.[59] This was because of his conviction that it was illogical to ask the Qatari people to fly the prescribed flag since Qatar was not a signatory to the Peace Treaty as Lorimer later believed.[60] In fact, Qatar, continued to use its own flag, a plain red flag featuring the name of the country, until the Ottoman landing in al-Bida in 1871.

The period between MacLeod's historic visit to al-Bida and the late 1820s was a quiet one for Qatar. The affairs of Qatar attracted little attention during this period except for an attempt by the deposed ruler of Abu Dhabi, Mohammad bin Shakbut, in exile in al-Bida, to regain his lost position. After the failure of this attempt he returned to Qatar and lived at al-Huwailah, north of al-Bida.[61]

CONCLUSION
Well before the arrival of the al-Khalifa emigrants, Zubara already featured fortified forts and towers. The al-Khalifa however

31

monopolised trade and attempted to increase their political power in the town, which brought conflict with the original settlers such as the al-Musallam. The emergence of Zubara as a free trade zone severely affected the trade of Basra, Muscat and the Persian coast, which led to frequent Persian raids on Zubara, although the free trade policy increased the town's importance as a commercial hub. The occupation of Bahrain by the Zubarans brought an end to the Persian raids on Zubara once and for all. This occupation was facilitated by the active support and co-operation of all the leading tribes of Qatar. The shifting of the al-Khalifa's activities from Zubara to Bahrain had no effect on the commerce and trade of the town. Although the decline of Zubara began in the 1790s because of the economic, political as well as the religious zeal of the Wahabis, it nonetheless became a power base for the Wahabis as well as the dethroned Omanis. The destruction of Zubara, together with its neighbouring town of Khor Hassan, brought an end to a powerful maritime force, who could threaten not only Oman but also any regional power. The ruin of Zubara was mainly due to foreign intrigues, conspiracy and wars. Nevertheless, the fall of Zubara and the flight of bin Jaber from Khor Hassan changed the course of the history of Qatar and contributed to the emergence of Qatar's most important town after Zubara and Khor Hassan: al-Bida.

Notes:

1. See the Translation of Instructions to Goncalo da Silveira, Portuguese Doc. no. 1, 12 September 1627, preserved in the National Council for Culture, Arts and Heritage Archive (NCCAHA), Doha.

2. *Lam al-Shihab, fi Sirat Muhammad b. Abd al-Wahab*, (hereafter cited as *Lam al-Shihab*), British Library, Ms. f. 235; see also Ahmad Mustafa Abu Hakima, History of Eastern Arabia: 1750-1800, Beirut: Khayat, 1965, p.67; see also H. R. P. Dickson, Kuwait and Her Neighbours, London: George Allen & Unwin Ltd., 1956, p. 26.

3.   See Kursun, op. cit., p. 33; for the al-Musallam's capital at al-Huwailah see John C. Wilkinson, Arabia's Frontiers: The Story of Britain's Boundary Drawing in the Desert, London and New York: I. B. Tauris, 1991, p. 42.

4.   Abu Hakima, op. cit., p. 67.

5.   Ibid., pp. 49-50; see also Dickson, op. cit., p. 26.

6.   Lorimer, op. cit., Geographical and Statistical, vol. IIB, p. 1952; see also Abu Hakima, op. cit., pp. 70-71.

7.   Abu Hakima, op. cit., fn. 5, pp. 70-71; see also Wilkinson, op. cit., pp. 41-43.

8.   Shaikh Jasim gave this information to Col. Francis Beville Prideaux, the British Political Agent in Bahrain, when the later paid a visit to the Shaikh at al-Wusail in 1905, in connection with the collection of data for Lorimer, who was engaged in writing his masterpiece, the *Gazetteer*; see the Genealogical Table of the Al-Thani (Maa'dhid) family of Qatar in Lorimer, op. cit., Historical vol. 3, Tables-Maps, Pocket, no. 8; see also Prideaux to the Resident in the Persian Gulf, no. 455, 23 December 1905, R/15/2/26. For the actual date of bin Thani's arrival at Qatar, see Haji Abdullah Williamson's 'Political Note on Qatar' in Colonel H. R. P. Dickson, Political Agent, Kuwait to the Political Resident in the Gulf, no. C-17, 18 January 1934, F. O. 371/17799.

9.   J. B. Kelly, Britain and the Persian Gulf: 1795-1880, Oxford: at the Clarendon Press, 1968, p. 26; see also Lorimer, op. cit., Geographical and Statistical, vol. IIB, p. 1534.

10.   See Samuel Manesty and Harford Jones, "Report on the Commerce of Arabia and Persia" Appendix F, in J. A. Saldanha, op. cit., Selections from State papers Bombay, regarding the East India Company's connections with the Persian Gulf, with summary of events, 1600-1800, vol. I, p. 408.

11.   Ibid., p. 408;

12.   Ibid., pp. 407-411.

13.   'Historical Sketch of the Uttoobee Tribe of Arabia (Bahrain)', in Bombay Selections, No. XXIV, p. 363.

14.   Ibid., p. 363; see also Kelly, op. cit., 1968, p. 26.

15.   Beaumont to William Hornby (President and Governor, Council at Bombay), no. 55, 15 July 1780, R/15/1/3; see also Bombay Selections, No. XXIV, p. 364.

16.   Galley to Hornby, no. 16, 5 October 1782, R/15/1/3.

17.   Bombay Selections, No. XXIV, p. 364; see also Galley to Hornby, no. 26, 7 June 1783, R/15/1/3.

18.   Galley to Hornby, no. 26, 7 June 1783, R/15/1/3; see also Bombay Selections, No. XXIV, p. 364.

19.   Lorimer, op. cit., Historical Vol. 1B, p. 840.

20.   Ibid., p. 840; see also Kelly, op. cit., p. 27.

21.   Lorimer, op. cit., Historical vol. IB, p. 789.

22.   This number was calculated later, by the Ottomans; see Sayid Mohammad Effendi (Mudir of Uqair and Agent of the Mudir of Zubara) to Commander Pelly, (HMS Sphinx), unnumbered, 26 July 1895, R/15/1/314.

23.   Lorimer, op. cit., Historical Vol. IB, p. 789; see also Kelly, op. cit., p. 27.

24.   J. A. Saldanha, op. cit., Vol. I, pp. 405-408; see also Abu Hakima, op. cit., pp. 176-178. For more details on Wahabi Movement, see also Lorimer, op. cit., Historical vol. IB, pp. 1051-1057.

25.   Husain Ibn Ghannam, *Kitab al-Ghazwat al-Bayaniyya Wal-Futahat al Rabbaniyya wa Dhikhr al-Sabab alladhi hamal ala*

dhalill, vol.-II, Bombay: (name of publisher is unknown), 1919, p. 161.

26. *Lam al-Shihab* in narrating the Wahabi raids on Zubara, stated that it was one of the richest ports in the Gulf, where the wealthiest Arab notables and merchants like Ibn Rizk and Bakr Lulu as well as some al-Khalifa merchants were living and dominating the trade and commerce of that part of the Gulf region, see Abu Hakima, op. cit., pp.154-156.

27. *Lam al-Shihab*, ff. 94-103; see also Abu Hakima, op. cit., pp. 158-159.

28. Abu Hakima, op. cit., p. 160.

29. 'Historical Sketch of the Utoobbee Tribe of Arabia (Bahrain)', Bombay Selections, No. XXIV, p. 366.

30. See Prideaux to the Political Resident in the Gulf, no. 208, 28 June 1905, R/15/2/26.

31. Lorimer, op. cit., Historical, vol. IB, p. 1075, 'Historical Sketch of the Rise and Progress of the government of Muskat' Bombay Selections, No. XXIV, p. 176.

32. Francis Warden, 'Historical Sketch of the Rise and Progress of the government of Muskat', Bombay Selections, No. XXIV, pp. 180-181.

33. Charles E. Davies, The Blood – Red Arab Flag, Exeter (U. K.): University of Exeter press, 1997, p. 327. In March 1810, when two Bombay Marine cruisers visited Khor Hassan, Abdullah bin Ufaysan was then residing at Zubara, see Kelly, op. cit., fn. 1, p.122.

34. 'Historical Sketch of the Rise and Progress of the government of Muskat', Bombay Selections, No. XXIV, pp. 182-183.

35. Ibid., p. 183; see also Lorimer, op. cit., Historical vol. IB, p. 843; see also Davies, op. cit., p. 327.

36.   Lorimer, op. cit., Historical vol. IB, p. 791.

37.   Kelly, op. cit., p. 28.

38.   James, Horsburgh, The India Directory, London: WMH Allen & Co., 1855, p. 422.

39.   Lorimer, pp. cit., Historical, vol. IB, pp. 789-790.

40.   Secret & Political Department, political consultation, Vol. 10, 6 January 1810, p. 208, Bombay Maharashtra Archive.

41.   Secret & Political Department, political consultation, Folio, 1050, 17 February 1810, pp. 209-210, Bombay Maharashtra Archive. See also Kelly, op. cit., p. 122.

42.   See ibid, Folio 1053, p. 212, Bombay Maharashtra Archive; see also Kelly, op. cit., p. 122.

43.   Secret & Political Department, political consultation, Folio, 1055, 17 February 1810, Bombay Maharashtra Archive.

44.   Ibid.

45.   Kelly, op. cit., pp. 122-123.

46.   Ibid., p. 123.

47.   Lorimer, op. cit., Historical vol. IB, pp. 791-792. See also Warden, SGB to John Adam, Secretary in the Secret and Political Department, unnumbered, 31 December 1816, R/15/1/19.

48.   Lorimer, op. cit., Historical, vol. IB, p. 792 ; see also J. Henderson, Secretary to the government of Bombay, unnumbered, 14 March 1817, R/15/1/19.

49.   Bruce to Warden, unnumbered, 29 June 1817, R/15/1/20. See also Lorimer, op. cit., Historical, vol. IB, p. 792.

50.   Warden, 'Sketch of the Proceeding of Rahma bin Jaubir, Chief of Khor Hassan', Bombay Selections, No. XXIV, p.

524; see also G. F. Sadlier to Sir Evan Nepean, President and governor in Council, Bombay, unnumbered, 17 July 1819, R/15/1/19.

51.   Bombay Selections, No. XXIV, p. 525.

52.   Bruce to the President and governor in Council, Bombay, unnumbered, 9 April 1820, R/15/1/22.

53.   For the full text of the Peace Treaty, see the Resident, Persian Gulf to the SGB, Bombay, No. 11, 7 February 1824, R/15/1/32.

54.   S. Hennell, Khor Hassan: continuation to the year 1831, Bombay Selections, No. XXIV, p. 528.

55.   For the full text of the Treaty, see Aitchison, op. cit., 1933, p. 245.

56.   Ibid., p. 232.

57.   See Article 3, in ibid., p. 245.

58.   Lorimer, op. cit., Historical vol. IB, p. 793; see also Charles Rathbone Low, History of the Indian Navy (1613-1863), vol. 1, London: Richard Bentley and Sons, 1877, p. 405.

59.   MacLeod to Warden, No. 8, 27 February 1823, R/15/1/30.

60.   Lorimer, op. cit., Historical, vol. IB, below p. 794.

61.   Rosemarie Said Zahlan, The Creation of Qatar, London: Routledge, 1989, p. 36.

Plate 1: Carsten Nieubuhr: an early cartographer of Qatar

Plate 2: The ruined eighteenth century (or earlier) settlement of Furaiha situated about two miles north of Zubara

Plate 3: *The destroyed old port of Zubara*

Plate 4: Ruins of Qalat al Murair built in 1778 and destroyed in 1811

Plate 5: Al Thaghab Well

The well is made of masonry six fathoms deep and still yields good drinking water

Plate 6: The destroyed fort of al Thaghab situated three miles south-east of Khor Hassan

CHAPTER THREE

The Struggle for Power in Bahrain and the Fall of Bin Tarif

The affairs of Qatar from the late 1820s to the 1840s were coloured by dynastic rivalries in Bahrain and the emergence of the coastal towns of Qatar as sanctuaries for those dislodged from Bahrain. Bahrain played one Qatari tribe off against another and involved itself in the affairs of al-Huwailah, forcing the Huwailah leader, Shaikh Isa bin Tarif, to go into exile in Abu Dhabi and later at Qais island, from where he continued his anti-Bahrain campaigns until he returned to al-Bida as its chief. Having established himself firmly at al-Bida, Isa bin Tarif and his deputy, Sultan bin Salamah, entered into conflict with the newly installed Bahraini chief. Eventually, a decisive battle took place near Fuwairet in which Isa bin Tarif was defeated and killed.

EXTERNAL PRESSURE ON BAHRAIN

It is crucial to examine the complex system of administration in Bahrain before discussing the history of Qatar during this period. When the second ruler of Bahrain Shaikh Ahmed bin Khalifa bin Mohammad, died around 1796, his two sons, Shaikh Salman and Shaikh Abdullah, became co-rulers of Bahrain; when Shaikh Salman died in 1825, his son Khalifa became co-ruler with Shaikh Abdullah. During the period of uncle-nephew joint rule, Qatar became a breeding ground for their political ambitions and a base for dissident factions of the al-Khalifa family. For example, in 1828, Shaikh Abdullah intervened in the Qatar peninsula when Mohammad bin Khamis, the step brother of Ali bin Nasir, head of the al-Bu Ainain tribe of al-Bida, stabbed an inhabitant of Bahrain. Shaikh Abdullah attacked al-Bida and destroyed the fort of the al-Bu Ainain tribe. When Mohammad bin Khamis was captured and imprisoned, his followers at al-Bida protested vigorously. Shaikh Abdullah retaliated by forcing them to move to Fuwairet, Ruwais and Zubara. Ali bin Nasir was forced to move to Wakra.[1] While Shaikh Abdullah involved himself in the affairs of Qatar, he soon faced serious external threats. Taking advantage of the weak administration of the Shaikh, the Muscat Sultan Sayyid Saeed renewed hostilities against Bahrain, occupying a Bahraini port and threatening an assault on the town of Manama towards the end of 1828. The threatened invasion of Manama was abandoned because of an outbreak of cholera on board the Sayyid's ships and the news of trouble in his possession in East Africa. Following an unsuccessful attack on the ships of Sayyid's subjects, Shaikh Abdullah eventually concluded a treaty with Sayyid in December 1829, agreeing to pay an annual tribute to Muscat in exchange for peace.[2]

The Wahabi Amir, Turki bin Abdullah al-Saud, posed a greater threat to Shaikh Abdullah than did Muscat. In 1830, Turki demanded that he should restore the fort of Dammam to Bashir, the son of bin Jaber, and pay Turki an annual tribute of 40,000 German Crowns. For want of an alternative, Shaikh Abdullah obliged, in return for Wahabi protection against external attack on Bahrain. Nevertheless, in 1833, Shaikh Abdullah repudiated Wahabi supremacy when Bashir abandoned his allegiance to the Wahabis and left for Oman from Tarut island opposite the town of Qatif. Taking advantage of the power vacuum created on the al-Hasa coast by Bashir's departure, Shaikh Abdullah, who became sole ruler of Bahrain upon the death of Shaikh Khalifa bin Salman on 31 May 1834, blocked the Wahabi ports of Qatif and Oqair with the support of the Amamarah section of the Beni Khaled. He occupied Tarut island and attacked the shipping of Qatif and Oqair. Faisal bin Turki, who became Wahabi Amir following the death of his father Turki in 1834, made an unsuccessful attempt to recover Tarut island and Dammam in 1835.[3]

BAHRAIN'S INTERFERENCE IN AL-HUWAILAH

Having established in hegemony on the al-Hasa coast, the old Bahraini ruler turned his eyes towards the Qatar coast, perhaps to divert the attention of the Bahraini people from his misrule. On 14 September 1835, Shaikh Abdullah ordered the sons (Mohammad and Ali) of Shaikh Khalifa bin Salman to proceed to al-Huwailah in

40

order to curb the growing power of Isa bin Tarif, the chief of al-Huwailah, and his deputy, Sultan bin Salamah. Bin Tarif was also leader of the al-Bin Ali tribe. Shaikh Abdullah's position became precarious when two of his sons by a woman of the al-Bin Ali (bin Tarif's tribe) fled on the night of 16 September 1835, accompanied by 400 men from Muharraq, and proceeded to al-Huwailah to join bin Tarif and bin Salamah, who already had 2000 fighting men. Bin Tarif's position was further boosted when he received an assurance from the Wahabi Amir that he would support him in the event of a Bahraini attack on al-Huwailah.[4]

Beating the war drum, the Bahraini Shaikh arrived in the deserted town of Zubara on 3 October 1835, accompanied by some of his sons, 1000 fighting men, 100 camels and 50 horses, and a plentiful supply of food and water. Declaring a trade embargo around al-Huwailah, he warned the people of the town to desert bin Tarif and join him or face the terrible consequences of non-compliance.[5] Wahabi reinforcements reached al-Huwailah via al-Hasa, consisting of 40 horsemen and 200 infantry with dates and rice. Now Shaikh Abdullah moved to Fuwairet to encircle al-Huwailah by sea and land.[6] Realising the gravity of the situation, bin Tarif sent bin Salamah and Ahmed, one of the rebel sons of Shaikh Abdullah, to Muscat on 15 December, to solicit the aid of the Sultan of Muscat. Sultan Sayyid Saeed turned down the request stating that he and Shaikh Abdullah had enjoyed peaceful relations for the last two years. He did, however, offer to mediate between the two warring parties to restore peace.[7] Bin Salamah agreed, and Sayyid Saeed sent his son Sayyid Hilal to al-Huwailah. The Bahraini Shaikh and the chief of al-Huwailah eventually reached an agreement. It was agreed that each party should retain the advantages acquired during the conflict, but that al-Huwailah should be evacuated and dismantled and the inhabitants transferred to Bahrain. There they would be settled, their personal safety guaranteed by Sayyid Saeed. The agreement was however immediately violated by Shaikh Abdullah's nephews and other partisans, who induced members of the al-Bu Kuwara tribe of Fuwairet to attack al-Huwailah, killing a dependent of bin Tarif.[8]

As Shaikh Abdullah refused to make reparations, bin Tarif, accompanied by bin Salamah and 400 followers, moved to Abu Dhabi, from where he hoped to carry out hostilities against the trade and territories of Shaikh Abdullah with British permission. The British authorities however declined to grant such permission on the grounds that he had established himself 'in a friendly and neutral port'.[9] Bin Tarif's capacity to engage in hostilities by sea was severely restricted in 1836, when Qatar was brought under the 'operations of the Maritime Truce of 1835', which extended the restrictive sphere from Halul island to ten miles north of Ras Rakkan, the extreme northern tip of the Qatar peninsula.[10] Having restricted bin Tarif's movements, the British authorities in the Gulf focused their attention on al-Bida and Wakra, in view of the reported maritime warfare of one Jassim bin Jaber Raqraqi, an Abu Dhabi outlaw and his associates off the coast of these two vital Qatari towns. The British made the chiefs of al-Bida and Wakra responsible for the unlawful activities of Raqraqi and his followers, although they had nothing to do with Raqraqi's acts. The East India Company's ship *Amherst* was deployed off Wakra to observe proceedings and check for maritime warfare at sea. As the pressure continued, its chief Ali bin Nasir left the place and joined with bin Tarif at Abu Dhabi in March 1838.[11]

BIN TARIF

As bin Tarif's movements were restricted, he started to correspond with the Egyptian governor, Khurshid Pasha, who had already established his authority in Medina, suggesting that they join forces to attack Bahrain. The possession of Bahrain by Khurshid Pasha would increase Egyptian influence in the Gulf and jeopardize British interests. Captain T. Edmunds, Assistant Resident in the Gulf, was therefore deployed by the government of India to reconcile Shaikh Abdullah and bin Tarif, preventing bin Tarif from allying with Khurshid Pasha. Bin Tarif agreed to suspend hostilities against the chief of Bahrain during the pearl fishing season of 1839, despite the Bahraini chief's rejection of the desire of bin Tarif to return to al-Huwailah and live peacefully.[12] Meanwhile, Shaikh Abdullah agreed in May 1839 with Mohammed Effendi, the confidential agent of Khurshid Pasha and the Egyptian governor of al-Hasa that he would pay an annual tribute of 3000 German Crowns to Mohammad Ali Pasha, the Ottoman Viceroy in Egypt in return for Egypt's non-interference in the affairs of Bahrain and the north-west coast of Qatar.[13]

Shaikh Abdullah's submission to the Egyptian ruler, in spite of bin Tarif's cease-fire, surprised Hennell, who had believed the Bahraini chief was motivated by a desire to frustrate Khurshid Pasha's plans to take control of Bahrain with the help of bin Tarif, and to deprive the latter of al-Huwailah.[14] Shaikh Abdullah's *entente cordiale* had wide repercussions in Bahrain, where a large portion of his tribe expressed dissatisfaction and decided to quit the island. As opposition increased, Shaikh Abdullah moved to

Khor Hassan in June 1839 and asked Mohammed Effendi to attack one of the dissenting groups, the al-Bu Kuwara tribe of Fuwairet. The attack was however successfully repulsed, and the invaders were compelled to retreat.[15]

Developments on the north-west coast of Qatar caused bin Tarif to lose his patience. Having failed to obtain British protection and permission to settle at Kharg island where Britain had temporarily transferred its Residency from Bushire, bin Tarif wrote to Hennell in August 1839, expressing his desire to reside either in al-Bida or Wakra under British protection. Hennell had no intention of getting involved, but did write to both bin Tarif and Shaikh Abdullah urging them to settle their differences amicably.[16] Consequently, bin Tarif agreed to cease hostilities provided Shaikh Abdullah restored his estates and property at al-Huwailah and Bahrain.[17]

As Shaikh Abdullah failed to respond quickly to bin Tarif's reasonable claims, the latter personally met Hennell on board the *Clive* at Muscat Cave in December 1839 to plead his case against the Bahraini chief. The al-Huwailah chief, reiterating his claims against the Shaikh of Bahrain, stated that he wished to leave Abu Dhabi as soon as possible because of his differences with the Abu Dhabi ruler. The Resident replied that he was unable to help. However, he would exert his influence with the Sultan of Muscat to obtain permission for bin Tarif and his followers to settle on Bassadore elsewhere within Muscat's territories. Bin Tarif declined this offer, arguing that as his followers earned their living by pearl fishing, the domains of the Sultan of Muscat were too far away. He stated that al-Bida was the only suitable place to live under British protection. If Hennell agreed to this, bin Tarif and his followers would devote themselves to the British cause in the Gulf. Hennell however, rejected this idea, as it failed to reflect British policy of considering bin Tarif as a dependent of Bahrain. He informed bin Tarif that Shaikh Abdullah had agreed to restore his and his followers' property and possessions if they agreed to reside on any stretch of the Qatar coast that he might assign and become subservient to him. Bin Tarif declined, arguing that no agreement would be made with the Bahraini chief without a British guarantee. He then disclosed that Khurshid Pasha's agent had offered him sanctuary at Qatif, but that he would not accept it as this would disturb his friendly relations with the British. Bin Tarif added that the time was ripe for Britain to take over Bahrain with his support and that of the al-Bu Kuwara tribe, who had already deserted Bahrain and taken shelter on Qais island, because of the oppression they had suffered at the hands of the Bahraini chief. The entire population of Bahrain would also support such a move. This was probably a negotiating tactic by bin Tarif, who hoped to bring Hennell round to his way of thinking and achieve his objective in Qatar. The adroit Resident was unmoved and showed no sign of interest in bin Tarif's Bahrain formula. The negotiations thus ended without tangible result.[18]

As bin Tarif failed to achieve concessions, the Sultan of Muscat intervened on his behalf. On 10 December 1839, at a private interview with the Resident, Sayyid Saeed told him that if the government of India was unable to meet bin Tarif's request, he should be allowed to establish himself at al-Bida under his own authority and protection, that is, independently. The Resident was reluctant to agree and explained:

> ...the reasons which had principally led me to oppose the hostile views of Esa bin Tareef against Shaik Abdoollah, were firstly the circumstance that his so doing would promote the views of Khoorshid Pasha in his then projected attack on Bahrein, and secondly the anomalous state of affairs which would arise from his carrying on hostilities from the Territory of a power which professed to be at peace, and to entertain friendly relations with his antagonist – that the connexion now subsisting between the Bahrein Chiefs and Koorshid Pasha removed in a great measure the first objection, while if the Ali Ally effected an independent establishment for themselves upon the Gutter [Qatar] Coast then dispute with Shaik Abdoollah would resolve into a mere local quarrel in which I did not see any call for our interference.[19]

The British Resident was worried that Khurshid Pasha might take over Bahrain, and that bin Tarif might establish an independent authority in Qatar. Hennell's reasoning was satisfactory to Sultan Sayyid Saeed.

The Sultan soon changed his position however. The next day, he sent bin Tarif along with Mohammed bin Saeed, chief of al-Bu Kuwara tribe of Bahrain, who also allied with bin Tarif, to Hennell with a note. The note requested acceptance of one of three positions, namely: (i) to locate the al-Bin Ali tribe of bin Tarif on the coast of Qatar under the protection of the British government; (ii) to consent to the Sultan's sending a force to establish bin Tarif and his tribe in al-Bida and (iii) to guarantee the restitution of their property and the future good behaviour of Shaikh Abdullah if they settled on the Qatar coast. The Resident replied that the government of India was unable to guarantee the first or third proposal. Regarding the second proposition, Hennell stated that bin Tarif should wait until he received the opinion of the government of India on the subject. In forwarding bin Tarif's case to Bombay, Hennell stated that his position had become stronger in view of the Sultan's backing. He was worried that Khurshid Pasha would support both bin Tarif and the Sultan against Bahrain unless deterred by British threats. In that case, Britain would lose the support of bin Tarif and the Sultan, who were ready to assist Britain against the Egyptians.[20]

As no definite reply was given to his request, on 16 December 1839, bin Tarif met Hennell on board the *Clive* to tell the Resident that he wanted to get out of Abu Dhabi as soon as possible, because of the economic difficulties his tribe was facing there as well as his deteriorating relations with Shaikh Khalifa bin Shakhbout, the chief of Abu Dhabi. This time he wished to move along with his people to the deserted town of Wakra, which was well located for pearl fishing and also at a sufficient distance from Shaikh Abdullah, with whom bin Tarif would not enter into conflict as long as Abdullah abstained from any attack upon him. This was acceptable to Hennell because of the location of Wakra and its situation:

> The Port of Wukrah (which is situated on the Gutter [Qatar] Coast between Bedda [Bida] and Adeed [Odaid] was a few years since inhabited by the Al Boo Synew [al Bu Ainain] Tribe whose Shaik Ally bin Nasir is now with Esa bin Tareef. The inhabitants paid no allegiance to the Bahrein Chief, with whom they were generally on bad terms. The exact cause of their quitting their Town is uncertain, but I believe arose partly from its possessing only an open roadstead which afforded little shelter to their vessels and partly their dread of the enmity of Shaik Abdoollah bin Ahmed.[21]

45

The former inhabitants of Wakra thus owed allegiance to no-one and had poor relations with the chief of Bahrain. It was probably this strained relationship between Wakra and Bahrain that made the former attractive to bin Tarif. Hennell recommended bin Tarif's proposal to the government of India, pleading that the proposed arrangement would not only gratify the Sultan of Muscat but would also deprive the al-Bin Ali chief of an excuse to forge ties with Khurshid Pasha. Furthermore, the British policy of suppression of maritime warfare would be bolstered by the establishment at Wakra of a chief like bin Tarif, who had 'afforded the most satisfactory proofs that he promises not only the power, but the desire to suppress and punish all proceedings of a piratical character'.[22]

On 20 December 1839, bin Tarif again visited the Resident on board the *Clive*. He told Hennell he had no alternative but to quit Abu Dhabi because of the increased hostility of Shaikh Khalifa bin Shakhbout towards him. Because of the possibility that Shaikh Khalifa would ally with the Bahraini chief to harass him and his tribe in their proposed new home of Wakra, the al-Bin Ali tribe was unable to take up residence at Wakra unless the British government prevented Shaikh Abdullah and Shaikh Khalifa from committing aggression. As Hennell expressed unwillingness to assist the al-Bin Ali chief, the latter made it plain that in that case his tribe would settle on the Persian Qais island, the Shaikh of which had agreed to accept them warmly. Hennell raised no objection to this as the island fell within the boundary line laid down by the British government as the limit of Arab maritime hostilities and the migrants would be free of fear of molestation from either the Shaikh of Bahrain or the Beni Yas chief of Abu Dhabi. At the same time bin Tarif and his tribe would be precluded from aggression against Bahrain or Abu Dhabi. At bin Tarif's request, a letter was sent by Hennell to Shaikh Khalifa bin Shakhbout, asking him to allow bin Tarif and his followers to leave Abu Dhabi without hindrance or molestation.[23] Eventually, in February 1840, bin Tarif and his supporters moved to the Qais island, where they settled peacefully.[24]

BRITISH ACTION AGAINST AL-BIDA
In the absence of an effective chief like bin Tarif on the Qatar coast, al-Bida became a sanctuary for foreign unlawful elements, forcing Britain to intervene. In November 1839, for example, one Ghuleta another Abu Dhabi outlaw took shelter at al-Bida after attacking a Basra boat off Bandar Dillam on the Persian coast. A. H. Nott,

Commander of the *Clive*, proceeded to al-Bida to demand that Salemin bin Nasir al-Suwaidi, the chief of the Sudan tribe at al-Bida to surrender Ghuleta. Al-Suwaidi was also warned not to allow unlawful acts to be committed from the port of al-Bida. Al-Suwaidi obliged and imprisoned Ghuleta, who had in February 1840, attacked one of the boats belonging to al-Suwaidi himself and seized its goods.[25] In addition to this action, al-Suwaidi also arrested Raqraqi, along with his associates when they attacked a Ras al-Khaimah boat, which had arrived at al-Bida for commercial purposes.[26]

Although al-Suwaidi's bold and unprecedented action with regard to Raqraqi, was appreciated by the Resident Hennell stating that his conduct was satisfactory to the Government of India,[27] the Resident still held him responsible for alleged maritime warfare off the coast of al-Bida and fined him 300 German Crowns in February 1841, alerting the Company's war ships in the Gulf to attack al-Bida, if al-Suwaidi refused to comply.[28] The British squadron *Coote, Sesostris* and *Tigris* arrived off the town of al-Bida on the morning of 25 February 1841, commanded by Commodore G. B. Brucks, who immediately served an ultimatum to al-Suwaidi, asking him to meet the British demand and also deliver up the captured boat or face the consequences. The Sudan chief declined to do so, arguing that he had nothing to do with Raqraqi's alleged activities. Defending his position, al-Suwaidi wrote back:

> I have received your letter in which you demand from me 300 Dollars. This demand is not a just one, the business which caused its being made belongs to Rugragee [Raqraqi] who is not a subject of mine, but his place is with the Benyas and Sheikh Khuleefa bin Shakboot. Should it be proved that I have gained anything by Rugragee's [Raqraqi] acts of piracy I will pay for every dollar thereby gained, 20 dollars. It was a Rusul Khyma [Ras al-Khaimah] boat which Rugratee [Raqraqi] had committed this act, I sent and seized both his boat and the one he had plundered, but he himself escaped. In doing this I thought I was performing a good act which all would approve of. Rugragee's [Raqraqi] boat is here and at your disposal. When my men fell in with the two boats they were empty and deserted. Enquire into the business and do justice. If I am in

fault, enforce your demand, if I am right to not harm me.[29]

In other words, al-Suwaidi rejected the British demand for payment of fine and defended his position with regard to the alleged maritime warfare. However, al-Suwaidi's clarification was unacceptable to Brucks. On the morning of 26 February 1841, the squadron fired on al-Bida and hit a small fort about one hundred and twenty feet square, as well as nearby houses. As Brucks threatened further bombardment, al-Suwaidi met the British demands in full.[30]

BRITHS CONTACT WITH BIN THANI

Having restored order at al-Bida, the British turned their eyes towards the northern coast, particularly Fuwairet, where they anticipated the outlawed members of Raqraqi might take shelter. On 27 March 1841, the Bahraini chief, on behalf of Commodore Brucks, wrote to Shaikh Mohammad bin Thani, the chief of Fuwairet, asking him not to harbour unlawful elements or afford them protection. It is noteworthy that this was the first time that British sources referred to Shaikh Mohammad bin Thani, the leader of the Maadhid tribe as the chief of Fuwairet. Shaikh Mohammad bin Thani was born at Fuwairet, the settlement of which was intensified following the arrival of the Maadhid from Furaiha under the leadership of Thani bin Mohammad bin Thani bin Ali. Ever since their arrival at Fuwairet, they formed a tribal confederation with al-Bu Kuwara who were also of Beni Tamimi descent.[31]

It is difficult to ascertain Shaikh Mohammad bin Thani's relationship with Shaikh Abdullah, who acted as middleman between Commodore Brucks and Shaikh Mohammad bin Thani in the affairs of Fuwairet at Brucks' direction. No doubt that he maintained cordial relations with the Bahraini chief. However, more than 2,000 members of the al-Bu Kuwara tribe, left Fuwairet for the Persian island of Kharg in 1839, in protest against the suppression of al-Bu Kuwara tribe in Bahrain at the hands of Shaikh Abdullah.[32] Although, the exact date on which Shaikh Mohammad bin Thani became the chief of Fuwairet is not available, it could be said that he emerged as the leader of the Maadhid–al-Bu Kuwara tribal confederation at Fuwairet following the exile of the al-Bu Kuwara tribe in 1839. Shaikh Mohammad bin Thani later played a prominent role in the history of Qatar and unified the whole peninsula under his leadership, as discussed in the next chapter.

Meanwhile, the most serious power struggle to date had flared up on Bahrain, severely affecting the western coast of Qatar, particularly Khor Hassan. In late 1841, Khor Hassan was turned into a base for two dissatisfied members of the al-Khalifa – Shaikh Mohammad bin Khalifa bin Salman (Shaikh Abdullah's grand nephew and joint subordinate ruler of Bahrain) and Shaikh Abdullah bin Ahmed, who had been living at Khor Hassan since 1838. Shaikh Mohammad paid a visit to the Qatar coast and stirred up opposition to Shaikh Abdullah among the inhabitants of Khor Hassan. The old Shaikh's position became yet more precarious when his sons openly defied his authority. Chaos and confusion reigned in Bahrain, forcing Shaikh Abdullah to seek peace; when Khalid bin Saud, the ex-Wahabi Amir, visited Khor Hassan in 1842, a rapprochement was achieved between the two co-rulers. Shaikh Abdullah and Shaikh Mohammad agreed to exchange places. The former would stay at Khor Hassan and the later in Bahrain. Soon afterwards, however, Shaikh Abdullah quarrelled with Shaikh Mohammad bin Khalifa when the latter attempted to prevent the marriage of a young girl living in Muharraq town, the headquarters of Shaikh Abdullah, to Ahmed, a son of Shaikh Abdullah. The aged Shaikh left Khor Hassan in June 1842 and attacked Manama, the headquarters of Shaikh Mohammad bin Khalifa, and drove Shaikh Mohammad out of Bahrain; he fled to al-Hasa. Having accomplished his mission, Shaikh Abdullah crossed over to Qatar and gave up Khor Hassan to 'partial plunder'. He then moved to Zubara and began to build the town, which had been deserted since 1811.[33]

In an attempt to dislodge the old Shaikh from the western coast of Qatar, particularly Zubara, Shaikh Mohammad bin Khalifa with 30 soldiers had easily seized Murair, close to Zubara, during the former's absence from the coast in February 1843.[34] He then established himself at Fuwairet with the strong support of the Fuwairet people, headed by Shaikh Mohammad bin Thani.[35] In the following month Shaikh Mohammad bin Khalifa with a strong army recruited in Qatar, landed at Rifa, 6 miles in the interior of Bahrain, and faced Shaikh Abdullah's son Nasir.[36] Shaikh Mohammad's position was further strengthened when bin Tarif and Bashir bin Rahma, son of Rahma bin Jaber, joined the battle on his side.[37] The coalition soon attacked Shaikh Abdullah from all sides and on 2 April 1843, after heavy fighting, forced him to retire to Dammam, which was governed by his other son Mubarak at the time.[38] The occupation of Bahrain by Shaikh Mohammad bin Khalifa was facilitated by the active support of bin Tarif and the Qatari contingent.

BIN TARIF IN AL-BIDA

Although it was anticipated by Lieut. Arnold Burrowes Kemball, officiating Resident (April 1843 to December 1843) that bin Tarif would stay at Bahrain and gain supreme authority there because of his strong personality and support from a large number of people in Qatar as well as in Bahrain, he instead went directly to al-Bida in May 1843 to establish Shaikh Ali bin Khalifa, brother of Shaikh Mohammad bin Khalifa, as the in-charge of al-Bida and made the necessary arrangements to transfer his residence from Qais island to al-Bida. He forced Salemin bin Nasir al-Suwaidi and his people to leave al-Bida fort and settle in Lingah. Following the ejection of al-Suwaidi, Shaikh Ali bin Khalifa took possession of the fort, while al-Bu Ainain took possession of Wakra and bin Tarif returned to Qais on 11 June 1843.[39] As the pearl fishing season came to an end, in October 1843, bin Tarif, along with his entire retinue including his deputy bin Salamah, arrived in al-Bida, settling to the south of the town. His position was further strengthened by the arrival of some members of the Maadhid tribe. The al-Bu Kuwara tribe headed by Mohammad bin Saeed Bu Kuwara also joined with bin Tarif.[40] The settlement of all these tribes at al-Bida under bin Tarif's authority undoubtedly enhanced the importance of the place. However, Shaikh Mohammad bin Thani remained in Fuwairet as chief; he made no attempt to move to al-Bida and conducted his business independently, while maintaining friendly relations with the new ruler of Bahrain.[41]

The departure of Shaikh Ali bin Khalifa from al-Bida in July 1843, as a result of the resumption of anti-Bahraini activities by the deposed ruler, cleared the way for bin Tarif's assumption of full power in al-Bida. Kemball welcomed this changed atmosphere as it brought bin Tarif and his 'tribe within the Arabian side of the restrictive line'. Kemball was convinced that the new al-Bida chief would tackle all irregularities at sea with the utmost sincerity, something that had proved impossible during al-Suwaidi's chieftaincy.[42] While maintaining law and order at sea to the great satisfaction of the British authorities, bin Tarif continued his co-operation and friendly relations with the new regime in Bahrain. In November 1843, bin Tarif paid an official visit to Bahrain and offered his services to the cause of Shaikh Mohammad bin Khalifa against his rival Shaikh Abdullah.[43]

THE BATTLE OF FUWAIRET

The period from 1844 to 1846 was one of comparative peace and tranquillity in Qatar and saw co-operation between al-Bida chief and the ruler of Bahrain. Relations however turned sour in 1847. Bin Tarif became suspicious of Shaikh Mohammad bin Khalifa's motives for manoeuvring on the northwest coast of Qatar. His frequent visit and interference in the affairs of Fuwairet and his efforts to rebuild Zubara convinced bin Tarif that Shaikh Mohammad bin Khalifa wanted to bring strategically located Zubara under his direct control to prevent future invasions of Bahrain by any hostile power based there. Therefore, when the deposed ruler Shaikh Abdullah, having gained support from the Wahabis, threatened to occupy Bahrain, bin Tarif changed allegiance, openly declaring his unqualified support for Shaikh Abdullah. Bin Tarif's dramatic move surprised Shaikh Mohammad bin Khalifa and he took advantage of the visit of Captain William Lowe, Commander of the Company's naval squadron in the Gulf to Bahrain to tell him about this new development on 2 November 1847. Expressing his grave concern, the disturbed Bahraini chief told Lowe that bin Tarif in conjunction with Shaikh Abdullah who had been living at Qais island since his dethronement, planned to attack Bahrain. The Bahraini chief further stated that he already deployed four armed bughlahs at sea to meet the invaders.[44] The formation of this 'unexpected coalition' between the 'powerful and energetic Chief of Bidda' and his former 'deadly enemy' had indeed, greatly undermined the position of Shaikh Mohammad bin Khalifa in Bahrain.[45] Assuring the chief of Bahrain that no fighting by sea would be allowed, Lowe wrote to bin Tarif urging him to desist from aggressive acts at sea and stating that armed vessels found for use in war would be seized at once.[46]

Lowe's warning failed to produce tangible results. The coalition demanded that Shaikh Mohammad should 'restore the vessels and possessions' of his grand uncle Shaikh Abdullah, which he had seized previously. Far from meeting these demands, on 4 November 1847, the Bahraini chief, blaming bin Tarif for creating a warlike atmosphere, complained to the Resident:

> The Commodore on his arrival here received information with regard to sundry matters in which I am concerned, and of which Bin Tarif is the cause. The latter contemplated an act of perfidy – he sent to Abdoollah bin Ahmed, his son Moobaruk, and

the others who are with him, of the Beni Hajir tribe, his object being to invade my territories, and subject the people of Guttur [Qatar] who are my dependants to himself. With this views he launched all his boats.[(47)]

Both bin Tarif and his deputy bin Salamah retorted by denouncing Shaikh Mohammad bin Khalifa, claiming that he was making plans for war. Attacking the Bahraini chief for 'treachery' against the people of Qatar and making it clear that it was their responsibility to defend their territory, on 7 November 1847, they jointly wrote to Samuel Hennell:

> We write to inform you with regard to Mahomed bin Khuleefa and his brother Ali, who have acted treacherously towards us, in return for the good, which as you know we did them. They had launched six Buglahs, and two Buteels, when the Commodore arrived at Bahrein, and put a stop to their proceedings. He also wrote an interdictory letter to us, and we desisted (from all hostile preparations).
>
> Subsequently Ali bin Khuleefa put to sea with the Buglah Tuweela, and three boats, and cast anchor at Fowairut in Kutr [Qatar]. We know not, what his object may be, but you are aware that Haweyla [Huwailah] is our country, and we fear but he should take it. Such is the way in which they harass and annoy us in exchange for the services we rendered there.[(48)]

The above letter reveals the courage and boldness of both bin Tarif and bin Salamah and their determination to resist the Bahraini chief from taking over of al-Huwailah, which they considered as their 'country'.

Bin Tarif was determined to defend the territorial integrity of the vital north-west coast of Qatar against any Bahraini infiltration. In the first week of November 1847, he and Mubarak bin Abdullah bin Ahmed landed at Fuwairet with 400 men. Upon hearing this news, Shaikh Mohammad bin Khalifa ordered seven small boats and twenty Battils with twenty horses and camels to land at Zubara and march upon Fuwairet.[(49)] As the military build-up intensified, the British authorities in the Gulf alerted their naval force and dispatched several war vessels including *Elphinstone* to blockade the long and narrow entrance of the harbour of al-Bida.[50] On 13 November 1847, the Resident warned the two belligerent parties:

> Accordingly I have deemed it expedient to address both parties reminding them of the irregularity of their proceedings in having prepared to commence naval hostilities without a previous reference to me but at the same time of the present case, it is not my intention to interfere between them so long as their hostilities are confined to Bahrein and the Guttur [Qatar] Coast, but that any of their Vessels found cruizing in the Persian side of the Gulf or elsewhere without the restrictive line would be immediately seized by our Vessels of War.[51]

Britain was primarily focused on Gulf maritime security, while events on land were of far less concern. Letters of warning were however delivered to Shaikh Mohammad bin Khalifa, bin Tarif and Shaikh Abdullah bin Ahmed. At the same time, a squadron began patrolling the Qatar coast.[52]

The military build-up continued. Bahraini troops numbering about 500, under the command of Ali bin Khalifa, landed at Khor Shaqiq (now al-Khor), a place between al-Huwailah and al-Bida. They were supported by Ahmed al-Sudairi and bin Osman, governors of al-Hasa and Qatif respectively. Shaikh Mohammad also joined them with extra troops and logistic support. The forces of bin Tarif, including those of his ally Mubarak bin Ahmed, amounted to around 600 men.[53] On 17 November 1847, the decisive battle between the coalition commanded by bin Tarif and the Bahraini troops under the command of Shaikh Mohammad bin Khalifa took place near Fuwairet. The fighting was short lived, and bin Tarif and eighty of his men were killed, which led to the final defeat of the coalition forces.[54]

As soon as Shaikh Mohammad bin Khalifa won the battle, he addressed the people of al-Bida, granting a general amnesty. He showed no sign of leniency however, to the members of bin Tarif's tribe.[55] Nevertheless, the battle of Fuwairet was a turning point in the modern history of Qatar; it paved the way for the rise of the al-Thani, the present ruling family of Qatar.

Having won the battle of Fuwairet, Shaikh Mohammad bin Khalifa led his war vessels to al-Bida, the headquarters of the defeated party. The Bahraini chief attacked the town and demolished it entirely, moving almost all the inhabitants to Bahrain. After Shaikh Mohammad captured al-Bida, Shaikh Abdullah moved to Nabend, on the Persian coast, while his son Mubarak fled to Nejd with about 200 followers.[56] More than 250 prisoners of war, including a son of bin Tarif and bin Salamah, were allowed by the chief of Bahrain to settle on Qais island.[57] Bin Salamah and his family later returned to Bahrain. The Bahraini chief sent his brother Ali bin Khalifa to the fort of al-Bida as his representative or envoy at al-Bida. However, he exercised no administrative or judicial powers at al-Bida or over other parts of Qatar. The local chiefs were still responsible for internal and external affairs. In November 1848 for example, one Khalifa, a member of the Qubaisat tribe of Abu Dhabi, appealed directly to Shaikh Mohammad bin Thani, the chief of Fuwairet[58], to grant him asylum, having failed to obtain the same from al-Suwaidi of al-Bida. While Shaikh Mohammad bin Thani was responsible for the administration of his domain at Fuwairet, al-Suwaidi, who had returned to al-Bida along with his tribe from exile after bin Tarif's death, had been exercising authority in al-Bida[59], despite the presence of a Bahraini representative there. Similar circumstances pertained in the neighbouring town of Wakra, which was under the absolute authority of the al-Bu Ainain tribe headed by Ali bin Nasir.

CONCLUSION

The battle in Fuwairet was the culmination of Bahrain's long-term interference in and designs on the north-west coast of Qatar, particularly Zubara, al-Huwailah and Fuwairet. The battle was however also a turning point in the making of modern Qatar and a reflection of bin Tarif's straightforward policy with regard to Bahrain's pretension in Qatar. Both he and his deputy bin Salamah were determined to frustrate the Bahraini chief's designs in Fuwairet and al-Huwailah once and for all. Despite his high morale and the strong backing of the deposed ruler of Bahrain, Shaikh Abdullah bin Ahmed and his son, bin Tarif lost the battle and was killed by Shaikh Mohammad bin Khalifa's forces. Mohammad bin Khalifa's victory was facilitated by the support of the governors of al-Hasa and Qatif as well as strong logistical support and reinforcements from Bahrain. While Abdullah bin Ahmed extended his full support to bin Tarif and bin Salamah and furnished them with men and materials, no leading Qatari tribes participated in the battle in support of bin Tarif's side. While a Bahraini representative was installed at al-Bida, following the conclusion of the fighting, tribal leaders such as Shaikh Mohammad bin Thani, Salemin bin Nasir al-Suwaidi and Ali bin Nasir remained in charges in Fuwairet, al-Bida and Wakra respectively.

Notes

1. Hennell to Shaikh Tahnoon, chief of Abu Dhabi, unnumbered, 18 October 1828, R/15/1/38; see also Lorimer, op. cit., Historical, vol. IB, p. 794.

2. Lorimer, op. cit., Historical, vol. IA, p. 448.

3. Ibid., vol. IB, pp. 856-857.

4. Agent at Bahrain to the Resident, unnumbered, 11 October 1835, R/15/1/68.

5. Ibid; see also Agent at Bahrain to the Resident, unnumbered, 20 October 1835, R/15/1/68.

6. Agent at Bahrain to the Resident, unnumbered, 18 November 1835; see also Agent at Bahrain to the Resident, unnumbered, 1 December 1835, R/15/1/68.

7. Agent at Muscat to the Resident, unnumbered, 25 December 1835, R/15/1/68.

8. Lorimer, op. cit., Historical, vol. IB, pp.794-795.

9. Kemball, 'Utoobee Arabs (Bahrein): Further continuation of the preceding, to the year 1844' in Bombay Selections, no. XXIV, p. 385.

10. Lorimer, op. cit., Historical, vol. IB, pp. 798 & 860.

11. Commodore John Pepper to Hennell, unnumbered, 27 January 1837, R/15/1/70; for Ali bin Nasir's flight to Abu Dhabi, see Hennell to Willoughby, no. 20, 7 May 1838, R/15/1/76.

12. Hennell to the Secret Committee, Court of Directors, Bombay, unnumbered, 21 March 1839; see also Hennell to I. P. Willoughby, SGB, no. 15, 2 March 1839. See also Hennell to Captain Edmunds, Assistant Resident in the Gulf, unnumbered, 10 March 1839, R/15/1/71; see also Kelly, op. cit., pp. 362-363.

13.   Enclosure no. 1 in Hennell to Willoughby, SGB, no. 57, 30 May 1839, R/15/1/71.

14.   Hennell to the Secret Committee, Court of Directors, Bombay, unnumbered, 6 June 1839, R/15/1/71.

15.   Hennell to Colonel Campbell, Agent and Consul General at Egypt, no. 25, 26 October 1839, R/15/1/80.

16.   Hennell to L. R. Reid, Acting SGB, no. 90, 28 August 1839, R/15/1/84.

17.   Hennell to the Secret Committee, Court of Directors, Bombay, unnumbered, 7 September 1839, R/15/1/84.

18.   Hennell to Reid, Acting SGB, no. 128, 13 December 1839, R/15/1/84.

19.   Ibid.

20.   Ibid.

21.   Hennell to Reid, no. 133, 17 December 1839, R/15/1/84.

22.   Ibid.

23.   Hennell to Reid, no. 140, 27 December 1839; see also Hennell to the Secret Committee, Court of Directors, Bombay, unnumbered, 18 February 1840, R/15/1/84.

24.   Translation of a letter from Khojah Rooben, Native Agent at Muscat to Hennell, unnumbered, 11 March 1840, R/15/1/85.

25.   Hennell to A. H. Nott, Commanding H. C. Sloop of War; *Clive* and Senior Naval Officer in the Gulf, unnumbered, 7 February 1840; see also Hennell to Willoughby, SGB, no. 11, 12 February 1840; also see Hennell to Willoughby, no. 36, 5 April 1840, R/15/1/87.

26. Substance of a Letter from Mullah Hossain, Agent at Sharjah to Hennell, unnumbered, 28 October 1840, R/15/1/92.

27. For full text see the ranslation of a Letter from Hennell to al-Suwaidi unnumbered, 12 December 1840, R/15/1/92.

28. Hennell to Brucks, unnumbered, 12 February 1841, R/15/1/92; see also Hennell to Willoughby, no. 11, 12 February 1841, R/15/1/94.

29. See the Substance of Correspondence between Brucks and al-Suwaidi, 25 February 1841, Enclosures in Brucks to Captain Robert Oliver, Superintendent of Indian Navy, no. 44, 27 February 1841, R/15/1/95.

30. For more details on the bombardment of al-Bida, see Brucks to Oliver, no. 44, 27 February 1841, R/15/1/95.

31. See the Substance of Correspondence between Commodore Brucks and Shaikh Abdullah, unnumbered, 27 March 1841, R/15/1/91; for the Maadhid – al-Bu Kuwara confederation see Wilkinson, op. cit., p. 43; see also Mustafa Murad al-Dabagh, *Qatar–Madiha Wa Hadiruha*, Beirut: Dar al Taliha, 1961, p. 176.

32. Kemball, 'Further continuation of the proceeding to the year 1844', Bombay Selections, No. XXIV, p. 390.

33. Lorimer, op. cit., Historical, vol. IB, pp. 867-868.

34. See the Enclosure no. 3 in Henry Dundas Robertson, Officiating Political Resident to Willoughby, no. 99, 2 March 1843, R/15/1/99.

35. Enclosure no. 1, in Robertson to Willoughby, no. 104, 16 March 1843, R/15/1/99.

36. Enclosure no. 1, in Robertson to Willoughby, no. 140, 27 March 1843, R/15/1/99.

37. Robertson to Willoughby, no. 160, 5 April 1843, R/15/1/99.

38.   See Hajee Jassem, Agent at Bahrain to Robertson, no. 51, 10 April 1843, R/15/1/99. For more on Shaikh Abdulla's asylum at Dammam Fort near to Qatar coast, see Kemball to Willoughby, no. 197, 25 April 1843, R.15/1/99.

39.   Agent at Sharjah to Kemball, unnumbered, 20 June 1843, R/15/1/99.

40.   Kemball to Willoughby, no. 546, 7 December 1843, R/15/1/99. For more on Maa'dhid's arrival at al-Bida see P. Z. Cox to S. M. Fraser, SGI/FD, no. 332, 16 July 1905, R/15/2/26. For al-Bu Kuwara's settlement at al-Bida, see Kemball to Hennell, unnumbered, 17 May 1844, R/15/1/104.

41.   For more on Shaikh Mohammad bin Thnai's authority at Fuwairet, see Hennell to Willoughby, no. 420, 1 October 1845, R/15/1/105.

42.   Kemball to Willoughby, no. 546, 7 December 1843, R/15/1/99.

43.   J. S. Draper, Commander E. I. C. Sloop *Coote*, no. 5, 29 November 1843, R/15/1/100.

44.   Captain William Lowe, Commander, East India Company's Naval Squadron in the Gulf of Persia to Hennell, no. 20, 8 November 1847, R/15/1/110.

45.   Ibid, see also Lieut. H. F. Disbrowe, Conclusion to the Year 1853, Bombay Selections, No. XXIV, p. 416.

46.   Lowe to Hennell, no. 20, 8 November 1847, R/15/1/110.

47.   Mohammad bin Khalifa to Hennell, enclosure no. 2, 4 November 1847, in Hennell to A. Malet, SGB, no. 477, 11 November 1847, R/15/1/111.

48.   Sultan bin Salamah and Isa bin Tarif to Hennell, Enclosure no. 3, 7 November 1847 in ibid.

49. Hajee Jassem, Agent at Bahrain to the Resident, Enclosure no. 1 and 1 in Hennell to Malet, no. 504, 1 December 1847, R/15/1/111.

50. Hennell to Malet, no. 477, 11 November 1847, R/15/1/111.

51. Hennell to Lowe, no. 482, 13 November 1847, R/15/1/111.

52. Hennell to Lowe, no. 493, 29 November 1847, R/15/1/111.

53. For more on the mobilization of troops, see Hajee Jassem, Agent at Bahrain, Enclosures no. 1 and 2 in Hennell to Malet, no. 504, 1 December 1847, R/15/1/111,

54. Hennell to Malet, no. 504, 1 December 1847, R/15/1/111.

55. For more on the battle of Fuwairet, see appendix no. 1.

56. Lowe to Hennell, no. 26, 27 December 1847, R/15/1/114.

57. Commodore T. G. Carless to Commanding Squadron in the Gulf, no. 2, 6 September 1848, R/15/1/114.

58. For more on the chieftaincy of Shaikh Mohammad bin Thani in Fuwairet, see Hajee Jassem to Hennell, Enclosure no. 1, 26 November 1848 in Hennell to Malet, no. 5, 9 January 1849, R/15/1/117.

59. See Hennell to Malet, no. 406, 31 December 1849, R/15/1/117.

CHAPTER FOUR

Bin Thani and the Unification of Qatar

Doha emerged as a town of prime importance and as the cockpit of foreign rivalry following the arrival of Shaikh Mohammad bin Thani. The attention of all the regional powers as well as of Britain was focused on this hitherto unimportant and neglected place. Shaikh Mohammad bin Thani's settlement at Doha filled the vacuum created by the al-Bu Ainain tribe's move to Wakra and marked the beginning of the emergence of Qatar as a separate entity under his leadership. The Wahabi Amir's brief occupation of Doha, from where he threatened to invade Bahrain, strengthened the position of the new chief of Doha considerably and set the stage for his future leadership of the whole peninsula. His determination, with the support of all the leading chiefs of Qatar, to keep the al-Khalifas from entangling themselves in the affairs of Qatar, led the al-Khalifa chief to destroy Wakra and Doha. However, Qatar's abortive counter-attack on Bahrain under Shaikh Mohammad bin Thani's leadership compelled Colonel Lewis Pelly to intervene in the long drawn out Qatar-Bahrain dispute and sign agreements with both Shaikh Mohammad bin Thani and the Bahraini chief. The Anglo-Qatari Treaty of 1868 recognised Qatar as a separate entity, and completed the process of national unification under the leadership of Shaikh Mohammad bin Thani 'The Chief of all chiefs' of Qatar.

BIN THANI'S MOVE TO Al-BIDA
In the early 1850s, following the moving of Shaikh Mohammad bin Thani from Fuwairet to Doha, British relations with Qatar developed significantly and British local officials began to come

into direct contact with the local Qatari chiefs. The alleged maritime warfare of one Soheil bin Ateish of Abu Dhabi on the coast of Qatar, particularly those involving a Kuwaiti boat, formed the background to a prime example of such contact. The British Resident, Hennell, held the chiefs of Qatar responsible for bin Ateish's unlawful proceedings at sea. In a letter to the government of Bombay, Hennell sought authority to deal with the chiefs in their domains of Doha and Wakra directly in the absence of any central authority or leadership. The chiefs referred to here were Shaikh Mohammad bin Thani and Rashid bin Faddal of Doha and Wakra respectively. Shaikh Mohammad bin Thani was the leader of both the Maadhid and al-Bu Kuwara tribes, while Rashid bin Faddal was the chief of the al-Bu Ainain tribe of Wakra.[1]

The sudden appearance of the chief of Fuwairet, Shaikh Mohammad bin Thani's as also the chief of Doha in January 1851 is worthy of note. This was the first time that the British Resident had mentioned Shaikh Mohammad bin Thani as the chief of Doha.[2] Doha, it is to be remembered, was situated on the bay of al-Bida. A small town called 'Doha al-Shagir' (Little Doha) lay between the old Doha town and al-Bida; only 400 yards of open space separated al-Bida from Little Doha. In addition to the Maadhid and al-Bu Ainain, other tribes living in Doha at that time included the al-Mussallam, who had already lost prominence, al-Bin Ali (from Nejd), Sulaithi and Manai.[3] Pinning down the exact date Shaikh Mohammad bin Thani moved to Doha and his reasons for doing so is a difficult task. He may have done so in 1850; until the end of 1849 he was at Fuwairet. Shaikh Mohammad bin Thani's move to Doha was politically and economically motivated. The town was previously dominated by the al-Bu Ainain tribe until their removal to Wakra. Since then, Doha had lacked effective leadership. While retaining his chieftaincy of Fuwairet, Shaikh Mohammad bin Thani moved there with the aim of filling the political vacuum and playing a decisive role in the affairs of Doha and developing his pearl trade. By the end of 1851, Shaikh Mohammad bin Thani had emerged as the most important figure in the Qatar peninsula.

In 1851, Doha was the focal point of British concern. Britain's attention was focused on it more minutely than before, because of Shaikh Mohammad bin Thani's apparent failure to take action against bin Ateish. The British authorities in the Gulf were seriously concerned by reports that bin Ateish had landed at Doha after attacking and taking possession of a Kuwaiti boat off Fuwairet. Although there is no evidence that bin Ateish came to Doha, the British made Shaikh Mohammad bin Thani as well as

Rashid bin Faddal, responsible for failing to deal with bin Ateish. The British eventually imposed fines of 500 German Crowns on Doha and 200 each on Wakra and Fuwairet.[4] After all the fines were collected through Mohammad bin Khalifa in March 1851. Despite his earlier statement of his inability to deal with the chiefs concerned, the Bahraini ruler, under British direction, forced the inhabitants of Wakra and Fuwairet to relocate in al-Bida and Bahrain to prevent them from sheltering bin Ateish in future. Moreover, in al-Bida they would be under the watch of his brother Shaikh Ali bin Khalifa, and in Bahrain he would be able to keep a close eye on them, himself.[5]

WAHABI PLANS TO INVADE BAHRAIN

Wahabi designs on Bahrain also contributed to the emergence of Shaikh Mohammad bin Thani as the most prominent figure in Qatar. In February 1851, it was reported by Hennell that the Wahabi Amir, Faisal bin Turki, moved from his headquarters at Nejd to a place called Joodah, between Qatif and al-Hasa. He was accompanied by a number of Bedouins and the Amamarah tribe and intended to invade Bahrain and land on the Qatar coast. Hennell thought that the Bahraini exiles at Qais island, headed by Mubarak bin Ahmed, son of the former Bahraini chief, would also join Faisal's cause.[6] Stating that he would definitely come to Qatar, Faisal outlined the reasons for his displeasure:

> ... the Bahrein Sheiks have squandered their property on my subjects and given them horses in order to attach them to their own side while the excuse themselves from paying me tribute. I will now send back to their horses to Guttur [Qatar] and reply to Sheik Ally's speech. It is said that the latter Chief once remarked to a messenger from Fysul "We will not be on good terms with the Ameer [Amamarah] unless he relinquishes the 4000 Dollars that he annually demands from us, for we are not Jews that we should pay tribute, if he will give up this claim we will be one with him but if not we will oppose him in Gutter [Qatar]".[7]

Faisal's decision to mass troops on the coast of al-Hasa and invade Bahrain was thus due to the Bahraini Shaikh's refusal to pay him tribute. Faisal's position was further boosted when Mubarak bin Abdullah promised to join Faisal, with his followers and more than

200 boats, as soon as Faisal had arrived at al-Bida and Doha, the al-Khalifa's supply route. Mubarak further agreed that once they arrived at al-Bida and Doha, they would pay 10,000 German Crowns to Faisal or a fixed annual tribute, provided Faisal occupied Bahrain for Mubarak and his followers.[8]

The Bahraini Shaikh took preventive measures in view of this gathering storm. Declaring a state of emergency in early February 1851, Shaikh Mohammad prohibited his subjects from communicating with Qatif and Oqair and prohibited the boats of his subjects from visiting India so that he could requisition them in the even of war. He asked his brother Shaikh Ali bin Khalifa, who was in Manama at the time, to proceed to al-Bida at once to enlist the support of the Qatar tribes; he was worried that they might assist the invaders should an attack be made on Bahrain. Shaikh Mohammad lacked the money to finance the war, and to raise a strong fighting force.[9] In view of this precarious situation, the merchants of Bahrain persuaded the Bahraini chief to open negotiations with his Wahabi counterpart; he offered 2000 German Crowns and two horses in tribute for that year. Faisal declined the offer and insisted upon the delivery of 10 horses, 10 camels and 4000 German Crowns in return for peace. Faisal's non-compromising attitude alarmed Hennell, as war would seriously disturb the peace at sea. He came to the conclusion that the best way to maintain order was the destruction of al-Bida and the removal of all its inhabitants to Bahrain 'in the event of their fidelity not being assured on a war with Faisal'. The Bahraini Shaikh should, Hennell suggested, implement this plan before his arch rivals – the sons of the late ex-chief of Bahrain – and the other Utubi refugees who had taken shelter on the island of Qais occupied the 'excellent harbour' of al-Bida.[10]

In view of Hennell's backing and his planning for the destruction of al-Bida, Shaikh Mohammad bin Khalifa rejected Faisal's demand in April 1851, maintaining that 'he would rather consent to be driven from Bahrein than comply with such terms'. At the same time he despatched his uncle to al-Bida to inform Shaikh Ali of his intention to go to war against Faisal. Shaikh Ali was instructed to warn the people of al-Bida that he would send vessels to destroy al-Bida town and remove all the inhabitants if they betrayed him and joined with his enemy. While the people of al-Bida declined to relocate to Bahrain, they promised that they would defend al-Bida in the event of a Wahabi attack. The Bahraini Shaikh was supported by the merchants of the island with their lives and property, despite their conviction that any war would spell disaster for the poor people engaged in pearl fishing. Shaikh Mohammad's position was further boosted by the offer of military aid from Shaikh Saeed bin Tahnoon of Abu Dhabi, who also sent a boat to Shaikh Ali at al-Bida.[11]

Tension mounted in May 1851 when it became clear that Faisal's troops were already marching towards al-Bida. Hearing this news, Ali bin Khalifa summoned all the pearl divers to defend the port of al-Bida and requested help from Shaikh Saeed bin Tahnoon, ruler of Abu Dhabi.[12] A final attempt was made to negotiate peace with Faisal. Shaikh Ali bin Khalifa proceeded to Zakhnuniyah island, nearly opposite Faisal's encampment, on 2 May 1851. Shaikh Ali sent Rashid bin Khalifa to Faisal with a letter announcing his arrival and requesting him to send either his brother Jilawee or his son Abdullah to negotiate for peace. The Wahabi Amir, however, replied that he would listen to no proposal unless Shaikh Ali himself submitted to him personally. Shaikh Ali refused this outright:

> If you are desirous of peace we will for this year present you with two horses & two Camels and at the time of Zukat will pay you as usual 4,000 Crowns. If you pleased to make peace on these terms well and good otherwise there is no further necessity to wait longer for us. You can go where you like and we will be there before you.[13]

The Wahabi Amir's demand was not limited to the camels and horses or Zakat (tithe). He wished to possess Bahrain. Shaikh Ali went to Bahrain to consult with his brother Shaikh Mohammad bin Khalifa and to discover whether Faisal would march against al-Bida or return to Qatif from his present encampment.[14]

BIN THANI'S SUPPORT FOR FAISAL

Preparations were made to face the Wahabi Amir in the event of an attack on al-Bida. All available vessels, men and provisions were placed at the disposal of Shaikh Ali. Orders were issued to increase the number of vessels opposite Qatif to blockade the Qatif port. On 18 May 1851, however, Faisal had arrived at a spring called *Ereeq*, within two day's march of al-Bida. Shaikh Ali at once returned to al-Bida. He recalled all the divers to the town and sent Mohammad bin Ahmed, an inhabitant of Abu Dhabi, to Shaikh Saeed bin Tahnoon to inform him of Faisal's march to al-Bida and request the promised aid.[15] However, when Faisal and his advancing troops reached the vicinity of al-Bida, Shaikh Ali was rocked by the unexpected change of attitude of Shaikh Mohammad bin Thani and other leading Arabs of Doha and al-Bida, who extended whole hearted support to the Wahabi Amir.[16] On 8 June 1851, to the great surprise of Shaikh Ali bin Khalifa, a contingent from Doha under the leadership of Shaikh Mohammad bin Thani took possession of the unattended *Borj-el-Mah*, the tower which commanded the wells at Doha. *Borj-el-Mah* was situated closely to the south-west side of al-Bida fort, where Shaikh Ali had been staying. Shaikh Ali, accompanied by Hamdan bin Tahnoon, who had been sent by the ruler of Abu Dhabi with military provisions, escaped to Bahrain without confronting the Doha force. Faisal entered at al-Bida in triumph. All the inhabitants of Qatar, including those of Fuwairet, al-Bida, Doha and Wakra welcomed Faisal and expressed their willingness to support him. Faisal's position was further strengthened by the support of Mubarak bin Abdullah, who had arrived in Doha along with some of his followers. However, Faisal was displeased with Shaikh Mohammad bin Thani for not arresting Shaikh Ali and allowing him 'to go free'.[17]

The loss of al-Bida to Faisal and the consolidation of his position in that part of the Gulf was a severe blow to the ruler of Bahrain. He was worried that the Wahabi Amir would soon launch an attack on Bahrain in collaboration with all the anti-Bahraini forces, including the Shaikh's fugitive relatives on the island of Qais. Shaikh Mohammad bin Khalifa therefore sent a substantial naval force to blockade al-Bida, and despatched Shaikh Hamad bin Mohammad to Resident Hennell at Bushire to seek British intervention in the event of a Wahabi-Qatari joint invasion of Bahrain. The Bahraini emissary stated that the chief of Bahrain was ready to place Bahrain under British protection if the British prevented the Utubi fugitives at Qais and the sons of the former chief of Bahrain from joining Faisal and his new allies at al-Bida and Doha. The emissary also stated that the Bahraini chief was willing to pay 10,000 German Crowns annually to the fugitives on Qais island if they abandoned their hostile attitude towards Bahrain and refused to ally with Faisal. While Hennell agreed to mediate and put the financial proposal to the fugitives, he made it plain that it was beyond his power to prevent them from joining Faisal's camp and that the Bahraini chief must depend upon his own resources to repulse external aggression against Bahrain. The Bahraini chief's offer to place himself under British protection was unacceptable to Hennell, in view of the altered situation created by al-Bida coming into the possession of Faisal and his ally, Shaikh Mohammad bin Thani.[18]

As Shaikh Mohammad bin Khalifa expected, half of the followers of Shaikh Mubarak bin Ahmed had already settled at al-Bida town. Mubarak himself arrived on Qais island in the last week of June 1851, to wind up affairs and quit with the remainder of his people to settle permanently at al-Bida. The concentration of anti-Bahraini forces at al-Bida forced Lieutenant A. M. Chitty, in charge of the Indian Navy HCS *Tigris,* to meet Shaikh Mubarak on Qais island and attempt to convince him not to join Faisal. Their conversation convinced Chitty that preparations were being made at al-Bida for an expedition against Bahrain, involving not only Faisal's forces but also the warships of Shaikh Mubarak bin Ahmed.[19]

BRITISH INTERVENTION
Although previously the British authorities in the Gulf had declined to defend Bahrain in any way, they now changed their minds in view of Chitty's intelligence report as well as the assessment of the British Minister at the Court of Persia, Colonel Justin Sheil, that in the event of a joint Qatari-Wahabi attack on Bahrain, the island would become an Ottoman dominion or protectorate. Following Sheil's assessment and without waiting for a definite reply from the government of Bombay, on 1 July 1851, Hennell instructed Commodore J. P. Porter to assemble all the vessels of war including HC Sloop *Clive* and the *Tigris* immediately at Bahrain harbour.[20] Faisal was warned that he should refrain from invading Bahrain and that the British Squadron would intercept his war vessels.[21] Hennell also informed Porter that he had written to the chiefs of Doha and al-Bida clarifying the British position. The Resident added:

> I have written to the Chiefs of Doah and Bidda to the effect that although it was not the intention of the British Government to interfere in the quarrel between the two branches of the Uttoobee tribe so long as this was confined solely to themselves, still they cannot be permitted to place their maritime resources at the disposal of a foreign power, and therefore they are interdicted, under pain of incurring the severe displeasure of the British Government, from employing their vessels for the invasion of Bahrein.[22]

In short, Hennell was determined to prevent the Qais vessels from joining the coalition against Bahrain. The British warship *Tigris* was ordered to anchor between Fasht Dibel and Bahrain island to prevent the people of al-Bida and Doha from attacking Bahrain from that direction. To repulse any attack from the Qatif front, Shaikh Mohammad bin Khalifa had already sent 11 war vessels, containing between 700 and 800 men under the command of Shaikh Ali bin Khalifa, the deposed Bahraini representative at al-Bida, to blockade Qatif port.[23] The blockading party was soon attacked off the port of Qatif by a fleet of 18 Bughlas belonging to the Utubi fugitives at Qais. In the resulting engagement Mubarak bin Ahmed, Shaikh Rashid bin Abdullah and Bashir bin Rahma, along with 150 members of the Qais party, were killed.[24]

FAISAL'S WITHDRAWAL

The sudden demise of Faisal's powerful Qais ally, Mubarak, caused him to adopt a new strategy. Abandoning his aggressive designs on Bahrain, on 20 July 1851, the Wahabi Amir proposed to send his two brothers and his son to meet Shaikh Ali bin Khalifa on board the Bughlah and effect a reconciliation. The proposal was rejected outright by Shaikh Ali bin Khalifa. Faisal became desperate for peace and sent Ahmed bin Mohammad Sudairi, chief of al-Hasa, to mediate. When Sudairi proposed to meet Shaikh Ali bin Khalifa on board his Bughlah, the latter declined the proposal stating that 'if you are agreeable to ask nothing, come, but if you want anything, don't come'.[25] Shaikh Ali bin Khalifa thus showed no sign of compromise and hardened his attitude. His hand was strengthened by active British support and the deployment of their naval squadron off the coast of al-Bida to protect Bahrain from invasion by the allied forces stationed at al-Bida and Doha. The morale of the Bahrain forces led by Shaikh Ali bin Khalifa was also bolstered by the shutting up of the allied fleets in Qatif by a superior blockading force.[26]

Attempts to reach a peace settlement continued despite the abortive Sudairi mission. Shaikh Saeed bin Tahnoon of Abu Dhabi, who maintained cordial relations with both Shaikh Mohammad bin Khalifa and Shaikh Ali, agreed to act as mediator in the Wahabi-Bahraini dispute at the request of the Sultan of Muscat and Sultan bin Saqar, Shaikh of Sharjah. Both the rulers of Trucial Oman were worried that the Wahabi Amir's prolonged stay in Qatar was a threat to the security of their own countries and feared that he might advance towards the Trucial coast to collect Zakat. The mediator arrived in al-Bida in the third week of July 1851; by 25 July assisted by Sudairi, he was able to persuade both parties to sign a peace agreement. While Shaikh Ali agreed to pay Faisal 4,000 German Crowns annually as Zakat, the Wahabi Amir agreed to restore the fort of al-Bida to the former, to disassociate himself from the people of Qatar and not to interfere on behalf of the sons of Abdullah bin Ahmed. Shaikh Saeed himself proceeded to Bahrain to obtain Shaikh Mohammad bin Khalifa's ratification of the agreement. Shaikh Mohammad bin Thani, who had played a vital role in driving out Shaikh Ali, also gave his consent. Shaikh Rashid bin Faddal, the chief of Wakra, refused to do so and went to Fars on the Persian coast. Following the conclusion of the peace agreement, on 26 July 1851 Faisal left for al-Hasa. Bahrain lifted blockade of Qatif and the British withdrew their naval squadron from Bahrain harbour and the coast of al-Bida, thereby paving the way for the return of the pearl fishing boats of Qatar and Bahrain to the pearl banks of the Gulf.[27]

COLLAPSE OF THE PEACE TREATY

The Wahabi-Bahraini peace treaty was threatened in 1852, when Faisal rehabilitated the surviving sons of Abdullah bin Ahmed at Dammam opposite Bahrain. Shaikh Mohammad complained about Faisal to the newly appointed British Resident, Captain Arnold Burrowes Kemball, during an interview with him on 3 April 1852. He threatened the Wahabi Amir with non-payment of tribute until he removed the sons of the ex-Bahraini chief from strategic Dammam. The Bahraini chief was however, worried that Faisal might incite Saeed bin Tahnoon, the Bahraini refugees on Qais, the Muscat authorities and the inhabitants of the Qatar coast against Bahrain. He therefore wanted to know whether the British government would intervene on his behalf in the event of an attack on Bahrain by the Wahabi confederates. The Resident made it clear that the inadequacy of the British naval squadron in the Gulf ruled this out.[28] As Kemball refused to listen to the Shaikh's pleas, he decided to pay tribute to the Wahabi Amir. However, he ordered a number of vessels, a Bughlah and two Battils to proceed to al-Bida and Doha to keep a watchful eye on the state of affairs there. The Bahraini chief also wanted to remove from al-Bida and Doha individuals considered likely to join with the Wahabi Amir against Bahrain. He wished to achieve this objective by preventing the Qatari people from engaging in pearl fishing, that is, by imposing an economic blockade.[29] The Bahraini deployment off the coast of al-Bida and Doha continued until the end of 1852, in view of the anticipated march of the Wahabi troops to Oman.[(30)]

The planned Wahabi expedition under the command of Abdullah bin Faisal against Oman in January 1853 inflamed the situation in the Gulf and once again Qatar inspired anxiety in Bahrain. It was reported in February 1853 by Hajee Jassem, British Residency Agent at Bahrain, that Abdullah bin Faisal had already arrived at al-Hasa and would soon march via Qatar to Buraimi in Oman. Although the people of Qatar assured the Bahraini chief that they would not cooperate with the Wahabis if they crossed Qatar, Shaikh Ali, who was at that time in Muharraq, had written to the people living at Zubara, probably the Naim tribe, that in the event of Abdullah bin Faisal's launching an attack upon them they were free to oppose him or retire to al-Bida.[(31)]

On 6 February 1853, the Resident was informed that Abdullah bin Faisal was likely to occupy Qatar and Khor al-Odaid, situated on a deep bay formed by the coasts of Seer and Qatar, on his way to Oman. Sending his brother Shaikh Ali to Qatar as his deputy to resist any Wahabi attack with the collaboration of the Qatari people, the Bahraini chief sought British assistance to frustrate Abdullah bin Faisal's designs on Qatar and Khor al-Odaid.[(32)] The Resident viewed Wahabi pretensions rather differently from Shaikh Mohammad, the Bahraini ruler:

> With reference however to your proposal to cooperate with Government for the purpose of opposing the establishment of Ameer Fysul on the sea coast I am inclined to believe that your surmises are certainly premature if not altogether ill founded but in any case Adeed [Khor al-Udaid] is part of the territory of Sheikh Saeed ben Tahnoon the Chief of Aboothabee who is mindful of his obligations and will I feel assured sanction nothing opposed to his engagements with the Br. [British] Govt. for the maintenance of Maritime tranquillity; & moreover although Ameer Fysul may have determined to depute an Agent as in past years to reside temporarily in Oman I have no reason to apprehend that he meditates any act inconsistent with the friendly understanding so long existing between him and the Sirkar. On the other hand while the Sirkar is quite prepared to defend its own rights & interests as it would wish also its allies to

70

be from whatever quarter they may be invaded it
will assuredly deprecate on their part as it carefully
avoids itself any hasty measures calculated to
induce the evil it is wished to avert.[33]

In other words, Kemball doubted the Bahraini chief's claim about
Wahabi intentions. He was unwilling to become involved in the
long-standing Wahabi-Bahraini fray. In fact, the Wahabis did not
advance to Oman via Qatar. Eventually, on 9 May 1853, the
Bahraini chief made peace with the Wahabi Amir, once again
through Kemball's mediation, and agreed to abandon hostilities.[34]

It was not long however, before Wahabi-Bahraini relations
turned sour, again affecting the affairs of Qatar. As early as 1859,
the Bahraini chief stopped the payment of the agreed tribute of
4,000 German Crowns and instigated certain tribes of Qatar to
attack Wahabi subjects. This was a response to continuous Wahabi
support for the Bahraini fugitives in Dammam and their plan to
capture power on the island. Denouncing this breach of faith on the
part of the Bahraini chief, Wahabi Amir Abdullah bin Faisal bin
Turki threatened to attack Bahrain. As tension mounted, the British
despatched a naval ship off the coast of Dammam to prevent
attacks.[35] Thereupon, Abdullah bin Faisal bin Turki bitterly
accused the Resident of intervening on behalf of Bahrain and
reminded him that the people of Qatar were Wahabi 'subjects' and
that the whole of Bahrain belonged to the Wahabi Amir.[36] Sudairi
provided further evidence in support of Wahabi links with both
the Bahraini and the Qatari people:

> Know that Bahrein used to give tribute to Ameer
> Fysul and to his father before him for ages past,
> long before the Sirkar [English East India
> Company] came to these parts. In like manner the
> Guttur [Qatar] people are also subjects of Ben
> Saood and used to pay him tribute.[37]

The Wahabi Amir was determined to recover his tribute
from the chief of Bahrain despite British interference. In May 1860,
the crisis took a serious turn when Abdullah bin Faisal bin Turki
threatened to occupy the coast of Qatar until Bahrain handed over
his 'annual subsidy' of 4000 German Crowns.[38] As Abdullah bin
Faisal kept up the pressure on Mohammad bin Khalifa, and
showed no sign of compromise, on 31 May 1861, the latter signed
an important landmark treaty with the British government. Britain

recognized the independence of Bahrain and Shaikh Mohammad agreed to 'abstain from all maritime aggressions of every description, from the prosecution of war, piracy and slavery by sea' in return for British protection against any attack on Bahrain by sea. The treaty was ratified by the government of India in February 1862.[39]

Although the 1861 agreement strengthened al-Khalifa's position at home considerably, he lost control over the affairs of al-Bida, as the result of the emergence of Shaikh Mohammad bin Thani as a predominant figure. William Gifford Palgrave, a British traveller who visited Qatar in 1863, and met Shaikh Mohammad bin Thani; stated that he was the 'acknowledged governor' or ruler of the entire Qatar peninsula.[40] In fact, Shaikh Mohammad bin Thani had established himself as the chief or ruler of Qatar during the short-lived Wahabi occupation of Doha and al-Bida.[41]

THE PROBLEM OF ZAKAT

The Wahabis once again came seriously to threaten the security of Bahrain. In November 1861, the British bombarded Dammam and removed the Bahraini 'pretender'. The Wahabi Amir nonetheless continued his struggle for tribute and deployed agents, who were 'active in sowing dissension amongst the tribes subject to Bahrein on the Guttur coast'. As 'anarchy' still prevailed as a result of Wahabi interference, on 8 February 1862, Captain James Felix Jones, the Resident (July 1856 to April 1862), appealed to the Wahabi Amir to make a peace with the Bahraini chief.[42] This appeal, however, produced no fruitful result. The Wahabi Amir still wanted 8,000 German Crowns from Bahrain. Attempted mediation by Shaikh Mohammad bin Thani in March 1862, was unsuccessful.[43] As there was no sign of compromise over the tribute issue, Shaikh Mohammad bin Khalifa referred the whole matter directly to the Porte, to whom the Wahabi Amri himself was said to be a 'tributary'. The Bahraini chief expressed willingness to pay the Ottoman government provided it granted him authority over the coast from Qatif to Muscat:

> Should assistance arrive for Sheikh Mahomed bin Khuleefa from the Porte, Sheikh Mahomed (as I understand) is not disposed to make peace with H. H. Ameer Fysul; when I waited upon him (the Bahrein Chief), he told me he had written engaging to pay 1,00,000 Dollars to the Turkish Government provided he were vested with authority over the

entire coast extending from Kutiffe [Qatif] to Muskat, were furnished with assistance against Ameer Fysul and were protected from interfering on the part of others (meaning the English).[44]

Shaikh Mohammad bin Khalifa's desire to buy 'authority' over the coast from Qatif to Muscat demonstrates his desperation to establish control over the coast of Qatar and thus prevent attack on Bahrain from that direction. The Ottomans however declined his offer though they agreed to attempt to mediate in the dispute between the Shaikh and the Wahabi Amir.[45]

The mediation efforts of the Ottomans and the British produced no result. The British nonetheless still hoped to achieve a 'pacific solution' to the long-standing disputes between the Wahabi Amir and the chief of Bahrain. In May 1862, Captain Herbert Frederick Disbrowe, Officiating Political Resident (April to November 1862), launched a fresh initiative to settle the quarrel equitably and investigated the main reasons for the two parties' divergent views. He discovered that the 'non receipt of Zakar or religious tithe for a considerable period' was the main 'source of irritation to the Wahabi Ameer'. He was convinced that if the Bahraini chief disbursed this 'contribution with regularity, the Wahabee Amir would willing come to terms of peace and would cheerfully do his utmost to restrain his subjects and protégés from molesting Shaikh Mohammad bin Khalifa's dependents and from forming combinations inimical to the security of Bahrein'.[46] In short, the Bahraini chief would have to pay a premium to his Wahabi counterpart to ensure the safety and security of Bahrain.

Disbrowe's logical thinking, prompted the Bahraini chief to settle the issue in mid May 1862, in view of the growing danger to the security of Bahrain from the Dammam coast. It was agreed that an annual tribute of 4000 German Crowns, would be disbursed regularly to the Wahabis, while the Wahabi Amir would forgo arrears of 1200 German Crowns. The relations between the two chiefs now 'stood on these foundations'.[47]

BAHRAINI INTERFERENCE IN WAKRA

Having improved relations with the Wahabi Amir, the Bahraini chief turned his eyes towards Wakra. In April 1863, he forced certain individuals to leave Wakra 'where divers in debt, and disreputable characters used to collect and injure trade or disturb the peace'. He also captured Wakra chief Mohammad bin Saeed Bu Kuwara, and brought him to Bahrain in custody.[48] Mohammad bin

Saeed Bu Kuwara, it must be remembered, was originally a subordinate of the chief of Bahrain and resided there. However, in the early 1850s he fled to the Persian coast due to arbitrary treatment by Shaikh Mohammad bin Khalifa. In 1859, after mediation by the British Resident had helped established peace, he made Wakra his place of residence[49] and had been living there since then. Hajee Ahmed, the British Agent at Bahrain gave reasons for his removal from Wakra by the chief of Bahrain:

> That personage [Mohammad bin Saeed Bu Kuwara] had always some secret correspondence with Ameer Fysul – and also used to give asylum to every one who committed any fault or misdemeanour at sea (for which the Sheikhs were to be responsible). They therefore sent men and brought him to Bahrein where they still retain him – and since then the inhabitants of Wakra have all been dispersed & scattered about.[50]

In fact, the Bahrain Shaikh's actions in Wakra were primarily motivated by Mohammad bin Saeed Bu Kuwara's alleged link with the Wahabis. The British government endorsed his actions in November 1863.[51]

RESUMPTION OF TRIBUTE PAYMENT

In December 1865, Shaikh Mohammad bin Khalifa despatched an envoy to the Wahabi capital with the promised 4000 German Crowns tribute. Although the British had previously supported the payment of tribute to the Wahabis as it made it less likely that they would invade Bahrain, they now opposed it, as it would make the al-Khalifa subordinate to the Wahabis. Moreover, the payment went against Bahrain's declared policy of treating Bahrain as an independent state, following ratification of the Anglo-Bahraini Treaty of 1861. In view of this complication, Colonel Lewis Pelly, Resident in the Gulf (November 1862 to October 1872), held an extensive inquiry into the whole issue of tribute or Zakat.[52] In December 1866, Pelly made an important discovery:

> ...that the Sheikh of Bahrein holds himself absolutely independent of the Wahabee Ruler in so far as the Islands of Bahrein are concerned. But the Chief having occupied certain lands in Guttur [Qatar] on the mainland of Arabia receives revenue

from these lands and pays to the Wahabee Ameer the sum of (4,000) Four thousand Dollars [German Crowns] per annum, on condition that the latter prevent his tribes from molesting the people residing on the Guttur lands.[53]

Thus, the purpose of this payment was to protect the Bahraini chief's land properties in Qatar, from Wahabi occupation. However, the Wahabi Amir himself disputed the contention that Bahrain was independent, stating that 'God had given him all the lands of Arabia from Koweit round to Ras ul-Udd [al-Hud]' therefore, including not only Bahrain but the Muscat territories as well as the port of Kuwait. The Wahabi Amir thus laid claim to all of Bahrain and Qatar. Pelly however, declined to accept 'these pretensions'.[54]

On 18 December 1866, Pelly reported his findings on the problem of tribute to the governor general in Council, who in March 1867, reiterated the British position that the Bahraini chief was liable to the Wahabi Amir for tribute, on account of certain lands held on the Arabian coast that is in Qatar, and so far as such property was concerned, 'it may be presumed that he owes fealty to the Wahabee Government'. Therefore, the Council concurred that with regard to Bahrain proper the chief was independent of the Wahabis and 'owes allegiance to no other power'.[55] Later, the government of India investigated further the mode of tribute payment and the Qatari contribution to such payments hitherto. Pelly added that the total annual claim of the Wahabis on both Bahrain and Qatar was 20,000 Krans or 4,000 German Crowns. The annual Qatari contribution to this total was reckoned at 9,000 Krans (1800 German Crowns). He further stated that one year's tribute (9,000 Krans) was collected from the Qatari tribes by Shaikh Ali bin Khalifa. Of this, 4,000 Krans were paid to the Shaikh of the Naim tribe in Qatar and 5,000 Krans to the chief of Bahrain. Nothing was sent to the Wahabis. It was also confirmed that the chiefs of Bahrain in 'late times or any time levy no taxes whatever for their own benefit from Guttur' [Qatar].[56] Thus it can be said that the tribute was a kind of protection money to be paid to the Wahabis both by the al-Khalifas and the chiefs of the Qatar coast.

ATTACK ON WAKRA AND DOHA

While the governor general in Council had been re-examining the al-Khalifa-Wahabi relationship in the context of tribute payment, the al-Khalifa received a severe blow in Qatar. This happened in

June 1867, when Shaikh Ahmed bin Mohammad bin Salman al-Khalifa, who had been acting as the representative of Shaikh Mohammad on the Qatar coast since 1863, seized a Bedouin of the promontory and deported him to Bahrain. Shaikh Mohammad bin Thani and the chief of Wakra demanded his release. His refusal prompted them to expel him from Wakra (his headquarters); he was forced to take shelter at Khor Hassan. Hearing of this, the Shaikh of Bahrain released the prisoner at once and expressed his desire for peace and friendly relations. Shaikh Jasim bin Mohammad bin Thani, the eldest son of Shaikh Mohammad bin Thani, was invited to Bahrain to make a permanent peace. However, upon his arrival in Bahrain he was imprisoned. The Bahraini chief then despatched a large number of troops under the command of Shaikh Ali bin Khalifa and Ahmed bin Mohammad to punish the people of Wakra and Doha. They were soon joined by a contingent from Abu Dhabi. The main reason for the joining of Abu Dhabi with Bahrain was due to the conviction of the Abu Dhabi chief that al-Bida had always been the 'popular asylum for fugitives from Abu Dhabi'. In October 1867, the combined forces sacked the towns of Wakra, al-Bida and Doha 'with circumstances of peculiar barbarity'. According to Lorimer, 'the damage inflicted on the people of Qatar was over $200,000.[57]

As the Wahabi Amir had always maintained good relations with the Qataris, the victims immediately appealed to him for help. Taking up the Qatari cause, he threatened the Bahraini chief 'with hostilities if the booty were not returned, and the inhabitants restored to their home'. This threat, however, fell on deaf ears. As the Qatari tribes got no redress, eventually, in June 1868, they took the law into their own hands and made a retaliatory attack on Bahrain. A sea fighting ensued in which 60 ships were sunk and 1000 men killed. Consequently, the Bahraini chief, Shaikh Mohammad bin Khalifa released Shaikh Jasim in return for the release of the captured Bahrainis.[58]

As the Bahraini attack on Qatar by sea was contrary to the terms of the Anglo-Bahraini Treaty of 1861, and had disturbed the peace in the region, in August 1868 Pelly, decided to proceed with three war ships the *Clyde*, *Hugh Rose* and the steamer *Sind* to punish the offenders. Hearing of Pelly's manoeuvres, Shaikh Mohammad bin Khalifa fled Bahrain and took refuge at Khor Hassan. On 1 September 1868, Pelly arrived off the coast of Wakra, the major town of Qatar at the time. Most of the chiefs of the peninsula came on board and confessed that they had breached the maritime truce by launching expeditions against Bahrain, but complained to Pelly about 'the piratical plunderings or destructions of their property and towns on the part of the Bahrein Chiefs'. They expressed their willingness to sign any documents which might secure the general peace and to catch Mohammad bin Khalifa, who 'during a quarter of a century, was increasingly become the terror of his neighbours' and the main obstacle to all progress and accumulation of wealth in his own islands, blessed with abundant natural resources.[59]

Following the conclusion of his meeting with the Qatari chiefs, Pelly went to Bahrain on 6 September 1868, to extract reparation for the outrages. He at once deposed Shaikh Mohammad bin Khalifa and made Shaikh Ali bin Khalifa ruler forfeiting all the 'war buglahs' belonging to Shaikh Mohammad bin Khalifa, he forced the newly installed chief to pay a fine of one hundred thousand German Crowns for violation of the Treaty of 1861. Shaikh Ali further agreed to hand over Shaikh Mohammad bin Khalifa to the Resident if he returned to Bahrain.[60]

THE 1868 TREATY

Having settled the Bahraini side of the equation, Pelly once again arrived at the Qatari port of Wakra in HMS *Vigilant* and on 11 September 1868, addressed a letter to Shaikh Mohammad bin Thani, the principal chief of Qatar peninsula. The letter detailed his arrangement with the new Bahraini chief, explained why he had overthrow Shaikh Mohammad by force, and warned the chief of Qatar:

> I have to warn you that if you conspire with Shaikh Mahomed against Bahrain or again put to sea for the purpose of disturbing the peace it will be my duty to take measures for putting it beyond your power to do further mischief.
>
> British Indian subjects had suffered loss at your hands by the destruction or plunder of their mercantile craft. I invited you to come on board at once and settle these questions. You will have safe conduct while on board.[61]

The next day, the Resident obtained an undertaking from Shaikh Mohammad bin Thani. The terms of which were as follows:

> I Mahomed bin Sanee, of Guttar, do hereby solemnly bind myself, in the presence of the Lord,

to carry into effect the undermentioned terms agreed upon between me and Lieutenant-Colonel Pelly, Her Britannic Majesty's Political Resident, Persian Gulf.

1st – I promise to return to Dawka [Doha] and reside peaceably in that port.

2nd – I promise that on no pretence whatsoever will I at any time put to sea with hostile intentions, and, in the event of a dispute or misunderstanding arising, will invariably refer to the Resident.

3rd – I promise on no account to aid Mahomed bin Khalifeh, or in any way connect myself with him.

4th – If Mahomed bin Khalifeh falls into my hands, I promise to hand him over to the Resident.

5th – I promise to maintain towards Sheikh Ali bin Khalifeh, Chief of Bahrein, all the relations which heretofore subsisted between me and the Shaikh of Bahrein, and in the event of a difference of opinion arising as to any question, whether of money payment or other matter, the same is to be referred to the Resident.[62]

In an apparent move to strengthen the position of Shaikh Mohammad bin Thani, on 13 September 1868, the Resident addressed a further note to all the leading tribal chiefs of Qatar, informing them of the new arrangement with Shaikh Mohammad bin Thani and asking them to refrain from engaging in hostilities against him:

Be it known to all the Shaikhs and others on the Guttar Coast that Mahomed bin Sanee, of Guttar, is returning with his tribe to reside at his town of Dawka [Doha], and has bound himself to live peaceably there and not to molest any of his neighbouring tribes. It is therefore expected that all the Shaikhs and tribes of Guttar should not molest him or his tribesmen. If any one is found acting otherwise, or in any way breaking the peace at sea, he will be treated in the same manner as Shaikh Mahomed bin Khalifeh, of Bahrein, has been.

> The British Resident, in the most friendly but solemn manner, warns all of you that the English Government are determined to preserve the peace at sea in the Persian Gulf. This Government has nothing to ask from any of you, but it is resolved to foster trade and to prevent pirates and other disreputable characters from injuring honest men who seek to grow rich and make their wives and children comfortable.[63]

The Agreement of 1868 and Pelly's subsequent note to the Qatari tribal chiefs signified that Britain treated Qatar as a separate political entity distinct from any other country in the region and recognized Shaikh Mohammad bin Thani as the predominant chief or ruler of Qatar for the first time. Although Pelly said little to reveal the real meaning of the Agreement, Captain Francis Beville Prediaux, Political Agent in Bahrain (October 1904 to May 1909) later clarified the status of Qatar under the Agreement of 1868:

> At some time between 1851 and 1866 A. D. Sheikh Mahomed bin Thani was enabled to consolidate for himself – no doubt with the good offices of the Wahabis, to whom Sheikh Mahomed of Bahrein had made himself very objectionable – a compact little dominion containing the towns of Wakrah, Doha and Bida, the independence of which from Bahrein was practically established and ratified by the Government of India in 1868 A. D., when a formal agreement was first taken from Sheikh Mahomed bin Thani.[64]

Prediaux's statement clearly indicates the emergence of Qatar as an independent state. Britain's signing of the Agreement of 1868 with Shaikh Mohammad bin Thani, on behalf of all the leading tribal chiefs of Qatar implies that the al-Thani family had already emerged as the leading power in the Qatar peninsula. The 1868 Treaty was thus an important landmark in the history of Qatar.

On 12 April 1869, Khamis bin Juma al-Bu Kuwara and Saeed Majed bin Saeed Darwish from Qatar and Bahrain respectively arrived at Bushire Residency to sort out financial matters with Pelly as stipulated in the Agreement of 1868. It was agreed that the annual sum payable by Qatar was 9,000 Krans, of which 4,000 would be paid to Rashid bin Jaber, chief of the Naim tribe and 5,000 to 'the Chief of Bahrain for ultimate payment to the Wahabee Government'. In determining the amounts involved the Resident made it clear to his visitors:

> It is understood that this payment of tribute does not affect the independence of Guttur [Qatar] in relation to Bahrein. But is to be considered as a fixed contribution by Guttur towards a total sum payable by Bahrein and Guttur combined, in view to securing their frontiers from molestation by the Naim and Wahabee Bedouins, more particularly during the pearl diving season, when the tribes of Guttur and Bahrein occupied at sea, leave their homesteads comparatively unprotected.[65]

The payment of tribute by the ruler of Qatar was thus intended to ensure the security of his newly emerged independent state. Although the Qataris continued the payment for two years (1868-69), Bahrain failed to send this amount to the Wahabis, implying that it kept it for itself.[66]

UNIFICATION OF QATAR

Shaikh Mohammad bin Thani who emerged as 'the most influential man in the whole promontory'[67] in the words of Lorimer, soon took effective steps to establish tribal harmony and unify the whole country under his leadership. In an effort to strictly implement the terms of the Treaty, the Shaikh restored order at Wakra, allowing British Indian subjects to settle there. Furthermore, in order to lessen the likelihood that the tribes of Khor Hassan would collude with Bahrain and to encourage them to merge with other Qatari tribes, Shaikh Mohammad bin Thani persuaded them to migrate from Khor Hassan to Wakra. This was confirmed by Captain Sidney Smith, Assistant Political Resident, who visited Wakra on 3 April 1869 in HMS *Dryad* to review the post-treaty situation in Qatar.[68]

Having consolidated his position at home, Shaikh Mohammad bin Thani followed an independent foreign policy with regard to the neighbouring powers. While he maintained good relations with Shaikh Ali bin Khalifa, he did not hesitate to ask the Bahraini chief to prevent the al-Bin Ali tribe of Bahrain from 'ill doing and from seizing' Qatari boats in the pearl fishing banks. As the Bahraini chief failed to act, Shaikh Mohammad bin Thani referred the matter to Pelly, maintaining that it was his responsibility as the independent ruler of Qatar to inform him 'before anything has taken place because you understand matters'.[69] The British authorities in the Gulf greatly appreciated Shaikh Mohammad bin Thani's independent foreign policy, particularly when the latter opposed Wahabi Amir Abdullah bin Faisal's plan to send troops to Oman.[70] The following letter from Pelly to Shaikh Mohammad bin Thani is illuminating in this regard:

> I have received your friendly letter, and am happy to find that all is quiet in Guttur. The happiness and wealth of you and your people lies in your prosecuting peaceful trade, and wholly abstaining from raids, which injure both yourself and others.
>
> You are not a British subject but an Arab Chieftain in friendly communication with H. M's Indian Government. Continue to conduct your affairs peacefully and with common sense and you will always find me ready to befriend you.[71]

Meanwhile, a significant change took place in Bahrain when the deposed chief, Shaikh Mohammad bin Khalifa was allowed to return to Bahrain to bring him under the direct control of Shaikh Ali bin Khalifa. However, Shaikh Mohammad bin Khalifa soon got involved in a conspiracy against his brother Shaikh Ali bin Khalifa, forcing the latter to deport the former to Kuwait, from where he moved to Qatif and continued his struggle for power. Eventually, a civil war broke out in Bahrain, which culminated in the killing of Shaikh Ali in September 1869. After a short 'interregnum' Shaikh Isa bin Ali was installed as the chief of Bahrain at Pelly's direction.[72] In Qatar, there was no change at the helm. By 1870, however, as Shaikh Mohammad bin Thani became weak due to old age, his son Shaikh Jasim began to assist him in discharging state functions.

CONCLUSION

The affairs of Qatar and Bahrain were thus stabilised under British direction. Imperial Britain established supremacy over both Bahrain and Qatar through the above-mentioned treaties with these two strategically important countries. The 1868 Treaty was both the culmination of continuous foreign interferences and a result of the Abu Dhabi-Bahrain joint attack on Wakra, Doha and al-Bida. The Agreement of 1868 recognised for the first time the separate entity of Qatar and the authority and leadership of Shaikh Mohammad bin Thani throughout the peninsula. Although the system of administration under tribal heads remained supreme, Shaikh Mohammad bin Thani rose to prominence and became chief of all the tribal chiefs, well before the signing of the treaty. The brief Wahabi presence in Doha and al-Bida no doubt contributed to the emergence of Shaikh Mohammad bin Thani as the first ruler and strongman of Qatar. Shaikh Mohammad bin Thani's move from Fuwairet to Doha thus suggests his far sightedness, determination and spirit of adventure. It was in Doha that he laid the foundations of the al-Thani dynasty, which has produced a lineage of able and progressive rulers.

Notes

1. Hennell to Malet, no. N, 4 February 1851, R.15/1/125.

2. For more on Shaikh Mohammad bin Thani's settlement at Doha, see Hennell to Malet, no. J, 23 January 1851; see also Hennell to Malet, no. N, 4 February 1851, R/15/1/125.

3. For more on the location and the description of Doha, see Captain C. G. Constable and Lieut. A. W. Strife (eds.), The Persian Gulf Pilot, London: Admiralty Hydrographic Office, 1864, pp. 104-105; see also Lorimer, op. cit., Geographical and Statistical, vol. IIA, pp. 487-488.

4. For more details on Soheil bin Ateish's maritime warfare off the coast of Fuwairet, Doha and Wakra, see Hennell to Malet, no. 4, 6 January 1851; see also Hennell to Malet, no. J, 23 January 1851; also see Hennell to Malet, no. 38, 25 February 1851; Hennell to Commodore Porter, no. 57, 20 March 1851, R/15/1/125.

5. See Hennell to Malet, no. 60, 20 March 1851; see also Mohammad bin Khalifa to Hennell, enclosure no. 1, 26 March 1851 in Hennell to Malet, no. 92, 4 April 1851, R/15/1/125.

6.   Hennell to Malet, no. 46, 3 March 1851, R/15/1/125.

7.   Hajee Jassem, Agent at Bahrain to Hennell, enclosure no. 1, 13 February 1851, in Hennell to Malet, no. 46, 3 March 1851, R/15/1/125.

8.   Hajee Jassem to Hennell, enclosure no. 2, 27 February 1851, in Hennell to Malet, no. 46, R/15/1/125.

9.   Hajee Jassem to Hennell, enclosure no. 1, 13 February 1851, in Hennell to Malet, no. 46, R/15/1/125.

10. Hennell to Malet, no. 130, 3 May 1851, R/15/1/125.

11. Hajee Jassem to Hennell, enclosure no. 1, 26 April 1851, in Hennell to Malet, no. 130, 3 May 1851, R/15/1/125.

12. Hennell to Malet, no. 158, 27 May 1851, R/15/1/125.

13. Hajee Jassem to Hennell, enclosure no. 1, 12 May 1851, in Hennell to Malet, no. 158, 27 May 1851, R/15/1/125.

14. Ibid.

15. Hajee Jassem to Hennell, enclosure no. 2, 18 May 1851, in Hennell to Malet, no. 158, 27 May 1851, R/15/1/125.

16. Hennell to Malet, no. 203, 18 June 1851, R/15; see also Hennell to Major A. Hamerton, Consul and Agent in Muscat, no. 189, 6 June 1851, ibid.

17. Hajee Jassem, Agent at Bahrain to Hennell, enclosure no. 2, 8 June 1851, in Hennell to Malet, no. 203, 18 June 1851, R/15/1/125.

18. Hennell to Malet, no. 203, 18 June 1851, R/15/1/125.

19. Lieut. A. M. Chitty, in-charge of H. C. B. *Tigris* to Commodore J. P. Porter, Commanding Persian Gulf Squadron, unnumbered, 24 June 1851, R/15/1/128.

20. Hennell to Porter, no. 219, 1 July 1851, R/15/1/125.

21.   Hennell to Amir Faisal, Regent of Nejd, enclosure no. 3, 7 July 1851, in Hennell to Malet, no. 226, 7 July 1851, R/15/1/125.

22.   Hennell to Porter, no. 219, 1 July 1851, R/15/1/125.

23.   Porter to Hennell, no. 124, 21 July 1851, R/15/1/125.

24.   See copy of correspondence from J. Tronson, Commanding H. C. Brigantine 'Tigris' to Porter, enclosure no. 1, 10 July 1851, in Porter to Hennell, no. 124, R/15/1/125.

25.   Porter to Hennell, no. 124, 21 July 1851, R/15/1/125.

26.   Hennell to Porter, no. 245, 30 July 1851, R/15/1/128; see also Hennell to Malet no. 252A, 2 August 1851, R/15/1/125.

27.   For more details on the conclusion of the Peace Treaty between the Wahabi Amir and Bahrain; see Hajee Jassem to Hennell, enclosures nos. 2 and 3, 28 July and 2 August 1851, in Hennell to Malet, no. 257, 5 August 1851, R/15/1/125; see also Hennell to Porter, no. 253, 4 August 1851, R/15/1/125; also see Porter to Hennell, no. 133, 31 July 1851, R/15/1/128.

28.   Kemball to Malet, no. 75, 3 April 1852, R/15/1/130.

29.   Hajee Jassem to Kemball, enclosure no. 1, 22 April 1852, in Kemball to Malet, no. 124, 27 April 1852, R/15/1/130.

30.   Kemball to Hennell, no. 423, 1 December 1852, R/15/1/130.

31.   Translated Extract from letter from Hajee Jassem, to Kemball, enclosure no. 1, 24 January 1853 to 5 February 1853, in Kemball to Malet, no. 57, 11 February 1853, R/15/1/138.

32.   Mohammad bin Khalifa to Kemball, enclosure no. 2, 6 February 1853, in Kemball to Malet, no. 57, 11 February 1853, R/15/1/138.

33. Kemball to Mohammad bin Khalifa, enclosure no. 3, 10 February 1853, in Kemball to Malet, no. 57, 11 February 1853, R/15/1/138.

34. Kemball to Amir Faisal, enclosure no. 3, 8 January 1855, in Kemball to H. L. Anderson, SGB, no. 18, 13 January 1855, R/15/1/149.

35. James Felix Jones, the Resident in the Gulf to Anderson, no. 396, 26 September 1859, R/15/1/179; see also Hajee Jassem to Jones, unnumbered, 29 September 1859, ibid.

36. See Amir Faisal to Jones, unnumbered, 14 November 1859, R/15/1/179.

37. Ahmed bin Mohammad Sudairi to P. W. Tendall, unnumbered, 29 September 1859, R/15/1/179.

38. Senior Naval Officer, Persian Gulf Squadron to James Felix Jones, No. 76, 20 May 1860, R/15/1/179.

39. For full text of the Treaty, see Aitchison, op. cit., pp. 234-236.

40. Palgrave, op. cit., 1865, p. 232.

41. See Kemball to Commodore Robinson, Commanding Indian Naval Squadron in the Gulf, no. 380, 14 November 1854, R/15/1/141; see also Hajee Yakoob, British Agent at Sharjah to Kemball, enclosure no. 1, 2 July 1855, in Kemball to H. L. Anderson, no. 278, 3 September 1855, R/15/1/149.

42. Jones to Faisal, unnumbered, 8 February 1862, R/15/2/29; see also R/15/1/183; also see Lorimer, op. cit., Historical, vol. IB, p. 890.

43. Hajee Ibrahim (Acting Agent at Bahrain) to the Resident, unnumbered, 21 March 1862, R/15/2/29.

44. Hajee Ibrahim to H. F. Disbrowe, unnumbered, 6 April 1862, R/15/2/29.

45. Disbrowe to J. Shaw Stewart, no 34, 16 May 1862, R/15/1/183.

46. Disbrowe to Kemball, no 18, 8 May 1862, R/15/1/183.

47. Disbrowe to Stewart, no. 38, 19 May 1862, R/15/1/183; see also Disbrowe to Stewart, no. 56, 16 July 1862, R/15/1/183.

48. Pelly to Anderson, no. 74, 13 April 1863, R/15/1/183.

49. See Pelly to Anderson, no. 122, 4 October 1863, R/15/1/183; see also extract from a Report on the Uttoobees by Disbrowe, from 1 May 1858 to 31 May 1859, R/15/1/183.

50. See the translation of Hajee Ahmed's Report regarding the Arab Tribe Al Bu Kuwara, unnumbered, undated, R/15/1/183.

51. See the SGB to the Acting Resident, no. 1512, 14 May 1863, R/15/1/183; see also the Officiating SGB to the Acting Resident, no. 3750, 14 October 1863, R/15/1/183.

52. Pelly to the C. Gonne, SGB, no. 52, 17 December 1865, R/15/1/183. See also the SGI/FD, to the SGB no. 958, 24 September 1866, R/15/1/183. Also see Lorimer op. cit., Historical, vol. IB, p. 892.

53. Pelly to Gonne, no. 127, 18 December 1866, R/15/1/183; see also R/15/2/29.

54. Ibid.

55. See Appendix no. 3.

56. Ibid.

57. For more details, see Lorimer, op. cit., Historical, vol. IB, pp. 892-894; see also J. A. Saldanha, op. cit., Précis of Bahrain Affairs, 1854-1904, vol. IV, pp.13-14; for Abu Dhabi's joining with Bahrain against Qatar see Prideaux to Cox enclosure no. 208, 28 June 1905 in Cox to Fraser, no. 332, 16 July 1905, R/15/2/26.

58.   Lorimer, op. cit., Historical, vol. IB, p. 894; see also Saldanha, op. cit., Précis of Bahrain Affairs, 1873-1904, vol. Iv, p. 14; see Kelly, op. cit., p. 674; also see Yousof Ibrahim al-Abdullah, A Study of Qatari-British Relations 1914-1945, Doha: Orient Publishing & Translations, 1983, p. 19.

59.   Pelly to Gonne, SGB, no. 111, 25 September 1868, L/P&S/5/261.

60.   See the Declaration in ibid; see also Aitchison, op. cit., p. 193.

61.   For full Text, see 'Appendix 6' in Pelly to Gonne, no. 111, 25 September 1868, L/P&S/5/261.

62.   See the 'Agreement of the Chief of El-Kutr [Qatar] Engaging not to commit any Breach of the Maritime Peace, - 1868', in Aitchison, op. cit., p. 255.

63.   See 'Appendix 8' in Pelly to Gonne, no. 111, 25 September 1868, L/P&S/5/261.

64.   Prideaux to Cox, no. 208, 28 June 1905, R/15/2/26.

65.   Pelly to the SGB, enclosure no. 1, 12 April 1869, in Pelly to the Duke of Argyll, no. 15, 24 April 1869, L/P&S/9/15.

66.   See Appendix no. 2; also see Aitchison, op. cit., p. 193.

67.   Lorimer, op. cit., Historical, vol. IB, p. 801.

68.   Captain Sidney Smith to Pelly, enclosure no. 2, 3 April 1869, in Pelly to the Duke of Argyll, no. 19, 24 April 1869, L/P&S/9/15.

69.   Translation of a letter from Shaikh Mohammad bin Thani, chief of Qatar to Pelly, enclosure no. 4, 8 July 1869, in Pelly to the Duke of Argyll, no. 40, 16 August 1869, L/P&S/9/15.

70.   Translation of a letter from Shaikh Mohammad bin Thani to Pelly, enclosure no. 2, 5 August 1869, in ibid.

71.   Pelly to Shaikh Mohammad bin Thani, enclosure no. 3, undated, August 1869, in ibid.

72.   J. A. Saldanha, op. cit., Précis of Bahrain Affairs, 1854-1904, vol. IV, pp. 21-23.

Plate 7: Bushire: the headquarters of the British Resident in the Gulf
(Watercolour, Lt. Col. Charles Hamilton, Smith)

Plate 8: View of Muscat city and fort, 1811

CHAPTER FIVE

The Ottomans in Al-Bida and the Ascendancy of Shaikh Jasim

The Ottoman occupation of al-Bida was a component in Ahmed Sefik Medhat Pasha's robust foreign policy, which he inaugurated in 1869, when he became the governor of Baghdad. It was Medhat Pasha's grand strategy to bring the length of the Arabian shore of the Gulf under Ottoman suzerainty and check British expansionism. To materialise his design Medhat Pasha extended Ottoman jurisdiction as far as the Qatari town of al-Bida, an enterprise which was accomplished after he had established his authority, firmly at Nejd. While Shaikh Jasim bin Mohammad bin Thani welcomed the Ottoman landing, his father Shaikh Mohammad bin Thani was far less enthusiastic. No sooner had the Ottomans attempted to establish a hold on Zubara, than the Bahraini chief claimed it on historical grounds. The claims and counter-claims continued until, the British government issued a final ruling in 1875, ordering the Bahraini chief to keep away from Zubara and all other parts of Qatar. Shaikh Jasim emerged as the sole ruler of Qatar upon the death of his father. However, relations with the Ottoman authorities both at al-Bida and Nejd became hostile, which eventually led to open conflict near Wajbah, in which the Ottomans forces were defeated. Following this battle, Britain tried unsuccessfully to mediate between Shaikh Jasim and the Ottomans. However, the Ottomans themselves restored peace and order in al-Bida through the direct intervention of Sultan Abdul Hamid II.

NEJD EXPEDITION
Prior to the occupation of al-Bida, the Ottomans wished to move against Nejd. The long running dynastic rivalry in Nejd between Abdullah bin Faisal and his brother Saud bin Faisal provided Medhat Pasha with the opportunity to fulfil his objectives. Dislodged by Saud in 1870, Abdullah sought Ottoman assistance through Shaikh Abdullah bin Sabah, the Kuwaiti ruler, to subdue Nejd. The Ottoman response was quick and decisive. Having obtained authority from the Grand Vizir Ali Pasha in June 1871, Medhat Pasha sent an expedition under the command of the divisional general Nafiz Pasha. Abdullah bin Sabah also accompanied Nafiz Pasha with his 300 war vessels, his whole-hearted support for Ottoman expansionism spurred by his fear of Saud's ambition in Kuwait. Nafiz Pasha soon defeated Saud's forces and the authority of the Ottoman empire was established throughout al-Hasa province, including Qatif and Oqair. Although the Ottomans were unable to extend direct rule to Nejd, they later gained control of it by proxy through Mohammed bin Rashid of Jabal Shammar.[1]

LANDING AT AL-BIDA
Establishing themselves firmly on the al-Hasa coast, the Ottomans next turned their attention to Qatar, which became a base for hordes of Bedouins who began to harass the Ottoman troops from the southward. Accordingly, Abdullah bin Sabah was sent to al-Bida with a Kuwaiti boat to clear the way for an eventual Ottoman landing. Bin Sabah carried four Ottoman flags to Qatar. Shaikh Jasim bin Mohammad bin Thani, the eldest son of Shaikh Mohammad bin Thani and heir apparent accepted one of the flags and hoisted it above his own house. A second flag was given to Shaikh Mohammad bin Thani, the ruler of Qatar. While he continued to fly the Arab flag above his house, he sent the Ottoman flag to Wakra. The third flag was handed over to Ali bin Abdul Aziz, chief of Khor Shaqiq (now al-Khor), a town north of al-Bida, and the fourth flag was sent to Khor al-Odaid, a place further south of Wakra and bordering Abu Dhabi.[2]

The hoisting of the Ottoman flag in Qatar was a severe blow to British prestige in the peninsula. Reacting sharply Colonel C. Herbert, British consul general at Baghdad, wrote to Medhat Pasha stating that 'the Ottoman authorities with the expedition have acted contrary to Your Excellency's intentions and repeated assurances'.[3] While no immediate reply was made to Herbert's letter, Major Sidney Smith, Assistant Political Resident in the Gulf, arrived at al-Bida in the gunboat *Hugh Rose* on 19 July 1871. Shaikh Mohammad bin Thani informed Smith that the Ottoman flag had been hoisted at al-Bida on the orders of the Ottoman commander at al-Hasa, who sent the flag through the ruler of Kuwait, and that he was unable to refuse to hoist it since the Ottomans had supremacy on the land of Qatar. Shaikh Mohammad bin Thani added that he had received a proclamation from Sulaiman bin Zahar, the Arab chief of Basra, in which it was written that the Ottoman Sultan was to bring the countries between Basra and Muscat under his jurisdiction and that all Arab chiefs desiring the Ottoman flag would receive it, together with, Ottoman protection.[4] However, Smith's fact finding mission in al-Bida convinced him that the Qatari chief had hoisted the Ottoman flag willingly. He was worried that there would be considerable 'risk of some quarrel by sea being fastened' on Bahrain by Qatar and acting under Ottoman protection.[5] To the great uneasiness of the British, the Ottoman authorities at al-Hasa sent a detachment of 100 troops and a field gun to al-Bida under the command of Major Omer Bay, in December 1871. They did so at the invitation of Shaikh Jasim, to whom Shaikh Mohammad bin Thani had already transferred his power because of his old age.[6] Medhat Pasha himself gave an account of the landing at al-Bida:

> …The rebels who have taken up arms against Sa'ud b. Faisal and the government are backed up by Bahrain which supplies them with food and ammunition. As the distance between Qatar and Bahrain is negligible, goods shipped from there are easily transhipped to the coast of Qatar. Sa'ud b. Faisal and the survivors of the recent conflict among the Shaikhs of Ajman and al-Murrah, allies of the former, repaired to the environs of Qatar to obtain the supply of food, ammunition and other necessities and to exercise pressure on the Shaikhs who had pledged allegiance to the Ottoman government. These rebels seized by force those herds of livestock belonging to the Qatar people and cut off their supply of fresh water. It is true that people in Qatar are capable of not only retaliating but even taking Bahrain under their domination. But as there are families among the Qatar people who have matrimonial links with

them, the inclination was compulsory [sic]. Under the circumstances, Jasim and Mohammad, his father, reiterated over and over again the necessity of despatching troops to Qatar. Whereupon a well equipped battalion positioned at Hasa was sent under the command of Major Omer Bey in addition to the corvette 'Iskenderiyye' and the vessel 'Asur' that sailed to Qatar.[7]

The Ottoman military occupation of al-Bida was intended at least in part to stop the encroachment by the tribes in the north-west of Qatar, which were backed by Bahrain. Nevertheless, the Ottoman landing put a brake on the operation of the Anglo-Qatari Treaty of 1868, and completed the extension of the Ottoman power from Basra up to the eastern limit of the Qatar peninsula.

Following the conclusion of the Qatar campaign, Medhat Pasha introduced administrative reforms over an area comprising al-Hasa, Qatif, Qatar and Nejd under the name of 'Nejd Mutasarrifgi' (Nejd governorship). In January 1872, Qatar was designated as a Kaza (district) under the Sanjak of Nejd, which was under al-Hasa province; officials were appointed for each Kaza and administrative unit. Shaikh Jasim was appointed as the Qaim-maqam, without salary.[8] The Ottoman presence in al-Bida thus brought no significant change in the administration of the country. Tribal relations continued on the same footing as before and the members of the al-Thani family were still the principal players in local affairs.[9] In fact, Qatar was exempted from paying all taxes except for Zakat as there was no cultivatable land in the country.[10]

STRUGGLE FOR ZUBARA
While the Ottomans strengthened their position in al-Bida, Zubara's status was hotly disputed. On 16 August 1873, it was reported by Major Charles Grant, the first Assistant Political Resident to Colonel Edward Charles Ross, Political Resident (October 1872 to March 1891), that a detachment of some 100 men had embarked from Qatif to accompany the Ottoman officer Hossain Effendi to Zubara. Although the report was unfounded, the Bahraini chief Shaikh Isa bin Ali reacted sharply and told Grant that the Naim tribes living near Zubara was 'his subjects under treaty' and their chief had acknowledged this 'subjection' in Colonel Lewis Pelly's presence.[11] Grant was unable to reply to the Bahraini chief's claim. He referred the matter to Ross, stating that

96

'he had no means of forming an opinion on the claim advanced by the Bahrein chief to sovereignty over the Naim tribe, but from verbal information he inferred that any power exercised by Bahrein of late years over that tribe had been merely nominal, if it existed at all'.[12]

Although Ross was sympathetic to the Bahrain chief's claim to Zubara and believed that the Naim tribe there had not accepted the Otoman suzerainty, he endorsed the views of the governor general in Council, and wrote back to Grant on 28 August 1873, stating that the Bahraini chief 'had not the power if he wished to protect tribes' residing in Zubara and that the British government was unable to intervene on his behalf as 'his so called rights, were involved in uncertainty'. Following Ross's instruction, Grant warned Shaikh Isa that he should 'remain strictly neutral' and keep himself 'aloof from all complications on the mainland with the Turks, Wahabis, etc'.[13] Despite this warning, on 2 September 1873, Shaikh Isa, once again attempted to assert his sovereignty over Zubara arguing that the Naim tribe of Zubara had signed a treaty with him previously. Following the examination of the old treaties by Grant and at Ross's recommendation the government of India, on 17 December 1873, issued a ruling on Zubara instructing Ross that he should make it clear to the Bahraini chief that 'it was desirable that the Chief should, as far as practicable, abstain from interfering in complications on the mainland' of Qatar.[14] The government of India did this, in order to avoid any Ottoman involvement in the affairs of Zubara and to maintain maritime peace in the region.

Despite the government of India's indifferent attitude, Shaikh Isa continued his efforts to gain Zubara. In 1874, he again saw a direct threat to his position when Nasir bin Mubarak, who lost his rights in Bahrain due to Shaikh Isa's opposition, moved to the Qatar coast. Shaikh Isa thought that soon Nasir bin Mubarak, with the assistance of Shaikh Jasim and other Qatari tribes, might first attack Shaikh Isa's close ally, the Naim tribe at Zubara, as a prelude to the invasion of Bahrain. As the Naim had aided Shaikh Isa in all of his previous crises, he wanted to send reinforcements to help them before they were attacked by Nasir bin Mubarak and his allies.[15] Nevertheless, Shaikh Isa's show of love for the Naim met with strong opposition from both Ross, and the governor general in Council. Reiterating that the chief of Bahrain had no possessions in Qatar, on 10 December 1874 the governor general in Council concluded:

His Excellency in Council therefore considers that the Chief of Bahrein should not have been encouraged to despatch troops to the mainland for the reinforcements of his allies, the Naiem tribe. On the contrary he should be advised to rely for support on the assistance of the British Government, which will, if necessary, be given him either to repel attacks by sea or to frustrate a threatening movement from the mainland.

It should be clearly pointed out to the Chief that so long as he adheres to his treaty obligations, the British Government will protect him: but if such protection is to be accorded to him, he must not be the aggressor or undertake measures which will involve him in complications and which are considered inadvisable by the British Government.[16]

Following the decision of the governor general in Council, Ross made it clear to Shaikh Isa that he should keep away from the 'feuds on the mainland' of Qatar. If he refused to do so the British government would not guarantee him protection against attack by sea.[17] Shaikh Isa was not the kind of person to heed such a warning so quickly. He exchanged several letters with the British[18] over the issue and continued his efforts until the British government issued a final warning in May 1875. Explaining the grave consequences of his interference in the affairs of Zubara, the Resident warned:

It becomes necessary under orders of the Government of India to explain to you as clearly as possible that if you should decide to persist in a course opposed to the advice of the Government of India and are thereby involved in complications on the mainland the consequences will be upon yourself, and the British Government will hold themselves free to take such measures with respect to you as they may think necessary.[19]

Now Shaikh Isa had no alternative but to oblige. He promised to cease interference in the affairs of Zubara; but refused to accept the extinction of his so called rights there.[20]

Although Shaikh Isa abstained from interfering further in the affairs oof Zubara, he maintained contact with the Naim tribe. Indeed, he did not hesitate to employ one hundred Naim of Zubara in his army and send occasional gifts as well as financial assistance to other Naim resident there. This he did in order to use them as his tools in times of necessity and prevent them from forming any anti-Bahrain coalition.[21] As Bahrain's alliance with the Naim tribe at Zubara was inimical to the security of Qatar, Shaikh Jasim became anxious to deal with the Naim. In 1878, an opportunity came when the pro-Bahraini Naims attacked on some al-Bida people at sea, Shaikh Jasim took action against them forcing some of the Naims to emigrate to Bahrain.[22] Although Zubara was then uninhabited, its strategic importance remained.

EMERGENCE OF SHAIKH JASIM

Shaikh Jasim became more prominent upon the death of his father Shaikh Mohammad bin Thani in early 1878. In May 1879, the Ottoman government, which considered Shaikh Jasim as one of the supporters of the Ottoman cause in the region, sent Zayyid Pasha, Mutasarrif of Nejd, to al-Bida to reassure Shaikh Jasim of Ottoman support. In return for this moral gesture Shaikh Jasim paid 9,000 to 10,000 Krans as tax to the Ottoman government.[23] However, despite this strengthening of his position externally, he soon faced internal disorder. For example, in November 1879, the al-Bu Kuwara tribe, instigated by the Shaikh of Bahrain, moved from al-Bida to settle in Fuwairet, denouncing Shaikh Jasim's pro-Ottoman policy. In 1880, the Beni Hajir, a leading tribe of al-Hasa, captured a boat belonging to Wakra and made a series of raids on Doha in the same year. In 1881, they were joined by some of the Naim and in the same year, the Ajman tribe carried off 450 camels from Qatar.[24] The lack of Ottoman support for Shaikh Jasim, despite his requests, disturbed relations between the two.

There was no subsequent sign of improvement in the Ottoman-Qatar relationship. Shaikh Jasim's relation with the Ottomans deteriorated further due to the Ottoman failure to assist him in his planned expedition against Khor al-Odaid, which was regarded by the British as belonging to Abu Dhabi. In 1882, despite British opposition, Shaikh Jasim, expecting Ottoman assistance, set out from al-Bida in the direction of Khor al-Odaid. But for want of Ottoman help, Shaikh Jasim was forced to abandon his plan to occupy Khor al-Odaid, and returned home empty-handed.[25]

The Ottoman's surreptitious support for Mohammad bin Abdul Wahab, who made a plot at al-Ghuriyah, north of al-Bida,

99

against Shaikh Jasim in 1886, claiming overlord-ship of that place, further alienated Shaikh Jasim from the Ottoman camp. In fact, Mohammad bin Abdul Wahab, who was an Ottoman subject, tried unsuccessfully to supplant Shaikh Jasim in the Qaim-maqamship of Qatar.[26] To achieve this objective he corresponded with the Ottoman authorities at Nejd and personally visited Bahrain to win the support of the Bahraini chief.[27] Shaikh Jasim, complaining about the backing of Nazih Bey, Mutasarrif of Nejd for Abdul Wahab, tendered his resignation on 25 February 1886. This letter of resignation was addressed to the Nazih Bey,[28] who strongly endorsed it, arguing that Shaikh Jasim had been troubling both the Mutasarrif and the supreme council at Constantinople with his occasional threats of resignation. The Mutasarrif recommended that the Ottoman government should, appoint a government representative in Qatar without delay, since a 'country without a government looks like a house without an owner'.[29] In this connection it is important to note that there was already an Ottoman quarter master and a judge at the fort of al-Bida.

The Sublime Porte declined to support the Mutasarrif's call for Shaikh Jasim to be replaced. Accordingly, on 3 March 1886, the supreme council instructed the Wali of Basra to inform Nazih Bey that the council was unable to change the Qaim-maqamship of Qatar. He was ordered to leave Shaikh Jasim in his post for the time being. The Porte, in fact, was determined to keep Shaikh Jasim as the Qaim-maqam of Qatar.[30] He was a balancing factor against the British, who had established a predominant position in the Gulf. Nazih Bey had no alternative but to accept the decision. Eventually, Shaikh Jasim and Abdul Wahab were reconciled and renewed their friendship when the latter was appointed as the Shaikh of Darein, near Qatif, by the Ottoman government.[31]

Shaikh Jasim's relations with the Ottoman became bitter when the latter attempted to establish a customs house at al-Bida. Thinking that the best way to frustrate the Ottomans' designs was to reduce the importance of al-Bida, he retired to a place called Ras al-Zaayen in the desert. Shaikh Jasim declared that he had severed his links with al-Bida and was no longer responsible for the administration of the country, which would in future be 'first referred to God and then to the Turkish Government'.[32] However, Shaikh Jasim's tactical retreat to the desert created lawlessness in al-Bida. In July 1887, two British Indian Banian traders were attacked and wounded by the residents of al-Bida because of their monopolization of the trade there. Although, Shaikh Jasim's brother, Shaikh Ahmed bin Mohammad bin Thani, and his son Shaikh Khalifa, protected the Banians, the British blamed the Shaikh. With the help of the Shaikh of Bahrain, Edward Charles Ross, the Resident in the Gulf, confiscated pearls worth 20,000 Indian Rupees (Rs), belonging to Shaikh Jasim in Bahrain. All the British Indian residents at al-Bida were removed to Bahrain and the Assistant Resident himself arrived there to remind Shaikh Jasim of his responsibilities. Shaikh Jasim sent an emissary to Bahrain to negotiate and eventually, the case was settled by the payment of around Rs. 6,000 as compensation to the injured British Indian subjects.[33]

British extortion of money from Shaikh Jasim caused him to write to the Mutassarif of Nejd on 3 November 1887, stating that he was being punished by the British for his links with the Ottomans and his co-operation with them:

> Wishing you all the best, I hereby submit to your consideration the fact that God Almighty and the entire Muslim community can bear witness that your most loyal subject has always been amongst the obedient servants of the Ottoman State and far removed from all worldly grudges. The foreigners [the British] feel restless in the face of my loyalty and every now and then they display their animosity; just to give you an instance of this, the British recently extorted from me 8000 rupees. This money has not been returned yet. Now my property has been seized and 6000 rupees have been exacted. I have run an additional loss of 5000 rupees due to the said attachment and for having had to wait [which caused the interruption of my business]. Such are the injustices to which I am subjected. All these are because I am an Ottoman citizen and protected by the Ottoman State, which seems to be against the British. These events are obviously contrary to the interests of the Ottoman State. Considering that the Ottoman State cannot possibly connive at this situation, I should like to suggest that, with a view to having my rights protected, the plaintiffs should henceforth have recourse to Ottoman courts of justice and act according to the judgments that are to be pronounced therein. Moreover, I beg your Grace to

intervene in the recovery of my property forfeited for no reason at all. May the Ottoman State prosper for ever.[34]

The Mutasarrif informed Mehmed Hafiz Pasha, Wali of Basra, of Shaikh Jasim's grievances. Noting Shaikh Jasim's contribution to the Ottoman cause, the Wali forwarded the message to the Ottoman Ministry of the Interior on 1 December 1887. The Wali maintained:

> ...The cruel treatment of Shaikh Jasim is due to the Shaikh's loyalty to and affection for the Ottoman State and to the un-rewarded excellent services he has rendered as qaim-makam of Qatar ever since the capture of Nejd. Shaikh Jasim is being deterred from serving the Ottoman State by harsh treatments in order that he may pledge allegiance to the British. This is unbearable. Jasim is an Ottoman citizen and a government official. All disputes should be settled by Ottoman courts of justice. Even in grievances related to cases of default, the courts dispensing justice should be the aforementioned ones. Under the circumstances, resort every now and then to cruelties and extortions from British officials instead of referring the cases to the appropriate authorities is not compatible, nor is it reconciled with international laws and treaties.[35]

Mehmed Hafiz Pasha recommended that the necessary measures be taken to protect Shaikh Jasim's interests. Upon receipt of this plea on behalf of Shaikh Jasim, the Porte protested to the British Ambassador at Constantinople. At the same time the Wali was instructed to pay a visit to Qatar to carry out an on-the-spot investigation of the matter. However, on 12 January 1888, Mehmed Hafiz Pasha proposed to the Grand Vizir that before his visit to Qatar, Shaikh Jasim, Qaim-maqam of Qatar, should be awarded the title of Kapucibasi (Head of the Palace Door Keepers) for his faithful services and loyalty to the Ottoman empire.[36] Eventually, on 29 February 1888, Sultan Abdul Hamid II issued a decree, awarding Shaikh Jasim the title.[37] Mehmed Hafiz Pasha arrived in al-Bida at the beginning of March 1888, and conferred upon Shaikh Jasim the title of 'Kapucibasi'.[38]

Upon his return to Basra, Mehmed Hafiz Pasha submitted to the Grand Vizir on 11 March 1888, a report on his fact finding mission, which recommended the permanent deployment of an Ottoman vessel off the coast of Qatar and upgrading the garrison at the old fortress at al-Bida to accommodate one infantry battalion or 250 men. The Wali's report also explained the British confiscation of Shaikh Jasim's property as due to his breaking ties with the British authorities in the Gulf and his pro-Ottoman leanings. The report also touched on Zubara, stating that it should be developed with new settlement to generate revenue for the Ottoman treasury. Shaikh Jasim should be asked to allow all the neighbouring tribes to settle there.[39]

Well before the implementation of the Wali's recommendation, Qatar was again in turmoil. This was due to a raid on al-Bida by 250 Bedouins, from Abu Dhabi, in which 24 men were killed, including Ali, a son of Shaikh Jasim. In May 1888, Shaikh Jasim wanted Ottoman support to retaliate against the raiders and avenge the death of his son. This was not forthcoming, yet Shaikh Jasim launched his expedition, and reached as far as Liwah in Dhafrah of Abu Dhabi territory in January 1889, and destroyed date plantations.[40] There followed raids and counter-raids by the forces of both sides, which continued until Shaikh Jasim's armed men penetrated into Sila, east of Khor al-Odaid, by boat in August 1889. Shaikh Jasim recalled his raiding party when the Resident, Ross warned him that Sila belonged to Abu Dhabi and he should withdraw his forces. Although Shaikh Jasim consented, he maintained that 'Sila was within Katar and therefore within his territory'.[41]

ADMINISTRATIVE REFORMS
Although the Ottomans extended no military or political backing to Shaikh Jasim's abortive Sila expedition, in June 1889 Akif Pasha, Mutasarrif of Nejd, submitted a memorandum to the newly appointed Wali of Basra, Hedayet Pasha, outlining administrative reforms to be introduced in Qatar. Akif Pasha's memorandum recommended the following reforms:

> The appointment to Zubarah, west of Qatar, of an able administrator conversant with the region to receive a salary of one thousand kurushes, under whose command there shall be some forty or fifty cavalrymen and infantry gendarmes. The appointment of an administrator conversant with the region, with a salary of seven hundred and fifty kurushes, who will reside at Odeid on the

demarcation line separating Oman and Qatar, south of Qatar under whose command there shall be a gendarmerie unit. The appointment of a harbour master conversant with the Arabic language, from the maritime arsenal for the control of ships carrying foreign passengers and Ottoman subjects and entering the large Qatar harbour and responsible for seeing that the harbour regulations are complied with. The appointment of an assistant to Jasim al-Thani, qaim-makam of Qatar, conversant with the Arabic language and political affairs with a salary of two thousand kurushes .... the elected administrative council members of Qatar, must regularly receive a minimum payment of two hundred and fifty kurushes on a monthly basis. The appointment of a correspondence clerk with a monthly salary of five hundred kurushes conversant with the Turkish and Arabic languages and with the correspondence forms and styles of both languages to take care of Qatar's correspondence, and the assignment of an assistant with a monthly salary of three hundred kurushes. These arrangements are sine qua non in view of the geopolitical situation in the region. Once these conditions are satisfied – by the Sultan's grace – the security of Qatar both from land and the sea will have been secured and a proper administration achieved. By our Sultan's grace again, two villages will be formed in no time in Zubarah and Odeid thanks to which commercial affairs and the pearl trade will thrive both at the said places and in Qatar. Thus, it will be possible in the future to realise an annual revenue of two thousand and five hundred liras derived from the ihtisap tax (market tax) like that which exists in Qatar and Ujair in addition to the harbour tax which is to be levied. The positive impact of these steps, which will create security and order both in here and along the other coasts of the Arabian peninsula, will evidently discourage foreign intrusions.[42]

These reforms were intended to bolster the Ottoman treasury, reviving the pearl trade from Zubara to Khor al-Odaid. The implementation of these reforms would also strengthen the Ottoman position along the Gulf coast up to Khor al-Odaid and prevent encroachment by foreign powers like Britain.

Akif Pasha's memorandum was sent to the Office of the Grand Vizir, on 9 October 1889, without any comment by Munir Pasha, Ottoman Minister of Interior. The proposals were discussed and approved by the Ottoman Council of Ministers on 26 December 1889; the Council emphasised that the proposed arrangements for Zubara and Khor al-Odaid should be implemented at once. On 2 February 1890, Sultan Abdul Hamid II ratified the decision and the Vilayet of Basra was instructed by the Ministry of the Interior to implement it.[43] The Porte's decision to establish administrative posts at Zubara and Khor al-Odaid became a cause of anxiety for the British government. The British considered Khor al-Odaid to lie within the British sphere of influence and if a settlement was established there under Ottoman direction it would be difficult to remove it by diplomatic means. Zubara, on the other hand, was situated on the western side of the northern extremity of the Qatar peninsula like a dagger pointing at the neck of the Bahrain island. The occupation of that island by the Ottomans would seriously jeopardize British interests in Bahrain.[44]

On 9 December 1890, the British Ambassador in Constantinople made the following query to the Ottoman government at the Foreign Office's instruction:

> Under instructions from Her Majesty's Government, Her Majesty's Embassy has the honour to request the Sublime Porte to be good enough to inform it as to the truth of a report which has reached Her Majesty's Government to the effect that the Imperial Ottoman Government intend to establish posts at Zobara and Odeid on the el- Katr [Qatar] Coast of the Persian Gulf, and that Mudirs have been nominated thereto.[45]

The Porte made no respond to the Ambassador's query with regard to Zubara and Khor al-Odaid. Therefore, on 14 March 1891, the British Foreign Office instructed the Ambassador to send a diplomatic note to the Ottoman government and made it clear that the British government 'could not acquiesce in the occupation' by the Ottomans of the two administrative posts mentioned above.

This was communicated to Said Pasha, the Ottoman Foreign Minister, on 22 March 1891 by Ambassador White stating that the chief of Abu Dhabi signed a maritime peace treaty with the Resident on 4 May 1853.[46] On 4 July 1891, the Sublime Porte responded, stating that Zubara and Khor al-Odaid were within the limits of the Vilayet of Basra and had long been administered by Qaim-maqam or Mudirs and that the Ottoman government had no knowledge of the Anglo-Abu Dhabi Maritime Agreement of 1853 and was unable to recognize it.[47] The British Foreign Office still remained unconvinced by the Ottoman clarification. The British Embassy at Constantinople wrote to the Porte on 26 August 1891, reiterating the text of the previous communication and enclosing the text of the agreement signed on 4 May 1853, between the British government and the Arabian chiefs including the chief of Abu Dhabi but excluding the chief of Qatar.[48]

The Anglo-Ottoman wrangling continued. On 26 January 1892, the Ottoman government replied to the British with a *note verbale*, maintaining that it had a legal right to establish administrative posts both at Zubara and Khor al-Odaid, and that the British Maritime Truce could in no way influence the Sultan's rights or sovereignty over the points in dispute. The note added that the Maritime Truce had nothing to do with the Ottoman plans for Zubara and Khor al-Odaid. The Porte warned that the British Political agent should not contact the Arabian chiefs on any matter without prior permission of the Porte.[49] In fact the British Maritime Truce of 1853, did not apply to Qatar as the country was not a signatory. However, the Porte was determined to preserve the Ottoman right to establish, military posts at Zubara and Khor al-Odaid to bring the coast of Qatar under its direct control, thereby restricting British movement in that part of the Gulf.[50]

Meanwhile, the situation in Qatar became precarious when it was reported that Nasir bin Mubarak, with the collaboration of Shaikh Jasim, planned to invade Bahrain from Zubara to dethrone Shaikh Isa. As this might disturb the maritime peace and order, on 17 August 1892 the Ottoman Council of Ministers decided to implement the reform programme previously envisaged for Qatar and send a battalion of soldiers to the region under the command of Mehmed Hafiz Pasha, Wali of Basra. It was thought that the deployment of troops would also prevent the Arabian desert tribes from launching more attacks on the trade caravans on their way from Oqair to al-Hasa.[51] The implementation of the Ottoman project in Qatar, that is, the stationing of administrators in Zubara and Khor al-Odaid, antagonised Shaikh Jasim who once again tendered his resignation from the post of Qaim-maqamship in protest. He also stopped paying taxes due to the Porte.

BATTLE OF WAJBAH

In October 1892, the Wali, Mehmed Hafiz Pasha, arrived in Nejd with his troops, consisting of 200 fighters.[52] He then led his forces to al-Bida, where he arrived with his troops in February 1893, to settle matters with Shaikh Jasim; in particular to seek payment of the unpaid taxes and confront him over his opposition to the proposed Ottoman administrative reforms.[53] Fearing that the Wali would kill or imprison him, Shaikh Jasim moved to Wajbah, about 12 miles south-west of al-Bida, accompanied by all the members of the Manasir as well as the Beni Hajir tribes. More than 400 members of other Qatari tribes also joined the Shaikh. The Wali sent a message to the Shaikh asking him to come and pledge loyalty. Shaikh Jasim sent one of his relatives, Shaikh Khaled, with a letter addressed to Mehmed Hafiz Pasha, stating that he was unable to see him unless the Wali withdrew his troops from Qatar. However, he was willing to pay Turkish Lira (TL) 10,000.00 in return for the withdrawal. Mehmed Hafiz Pasha declined the offer, stating that Shaikh Jasim must meet him, express his loyalty to the Ottoman government and disband his tribesmen.[54] Shaikh Jasim, thinking that the Wali's motive was to 'get him dead or alive', refused to come. The Shaikh took counter-measures, deploying part of his contingent at Salwa, cutting al-Bida garrison's land links with al-Hasa and seizing messengers as well as official correspondence, destined for al-Hasa. Consequently, the Wali was unable to contact the tribesmen who were coming to his aid under the command of Shaikh Mubarak al-Sabah, brother of the ruler of Kuwait.[55]

The Wali retaliated. On 25 March 1893, he arrested leading Qataris, including Shaikh Ahmed bin Mohammad bin Thani, Shaikh Jasim's younger brother, along with 13 leading men of al-Bida and imprisoned them on board the ship *Merrikh*, suspecting their involvement in the threatened Qatari attack. Declining to Shaikh Jasim's request to release them in return for ten thousand Liras, he despatched a column of troops consisting of 40 cavalrymen and 100 gendarmes with a cannon under the command of Yusuf Effendi to destroy Shaikh Jasim's fortress at Wajbah and seize the weapons stored there.[56] After two hours' march the Ottoman troops arrived at the fortress of Shebaka, a half an hour's march from Wajbah, and halted their first. No sooner had the Ottoman army resumed their movement towards Wajbah than they came under heavy gunfire from some 3,000 to 4,000 of Shaikh Jasim's infantry and cavalry armed with modern weapons. At night the Ottoman troops retreated to Shebaka fortress, where they were attacked by Shaikh Jasim's well-equipped forces. Demoralized by the lack of reinforcements and logistic support from al-Bida garrison and weakened by hunger and thirst, the Ottoman soldiers could not withstand the Qatari fighters. Eventually, they had no other alternative but to retreat to al-Bida garrison. As they retreated they were pursued by the victorious forces of Shaikh Jasim.[57]

Although a relief force from al-Bida rescued most of the surviving Ottoman forces, 11 had been killed and 55 were wounded in the battle. Moreover, 152 rifles and the 3-pound cannon were also captured by Shaikh Jasim's forces. The total loss on the Qatari side was estimated at 400 persons, including men, women and children. The Wali himself escaped to the Ottoman corvette *Merrikh*, and anchored at al-Bida harbour, with some of his soldiers, leaving behind 100 troops at al-Bida fort at Shaikh Jasim's mercy. The garrison came under heavy fire from Shaikh Jasim's advancing column, which forced the corvette to bombard the town. Nevertheless, Shaikh Jasim's forces besieged the garrison and cut the water supply in the neighbourhood of al-Bida. Their success forced the Wali to release Shaikh Ahmed and the other leading Qatari tribal leaders who had been holding as hostages. In return for their release, Shaikh Jasim granted the cavalry that the Wali had brought with him safe passage by land back to Hofuf. Eventually, the corvette with the Wali on board left the harbour and Shaikh Jasim continued to live at Wajbah peacefully.[58] The Ottoman Wali had miscalculated the strength of the Qatari troops and expected their surrender under pressure. To the great surprise of the Wali, the determined Qatari troops, under the command of Shaikh Jasim, not only repulsed the attack but also won the battle. Shaikh Jasim's victory in Wajbah was an important landmark in the history of Qatar. Following the conclusion of the battle, he not only emerged as the unrivalled hero of Qatar, but also the most important political figure in the region.

Having won victory in the battle of Wajbah, Shaikh Jasim decided to restore peace and order and improve his relations with the Ottomans. Therefore, on 27 March 1893, he wrote to the Grand

Vizir giving his version of the battle. The Qatari Shaikh complained:

> Despite unswerving loyalty and obedience to our Sultan, Hafiz Pasha, the Wali of Basrah, attacked us from land and sea, with the forces of the Imperial Ottoman army, without rhyme or reason. In return, we delivered the country along with its nahiyes to his hands and avoided all encounters. Upon his invitation, I sent him my brother. But he sent him along with other prominent members of this country to jail. This attitude caused hatred among the people and the tribes. I sent him word to the effect that even a governor of the state could not act against the orders of the state. Having disregarded our appeal, he wanted us to fulfil certain acts, which we did not fail to do. On the sixth day of the holy month of Ramadan, while we were in the desert, we were taken unawares. Certain tribes suffered losses, even women and children were not spared. In the face of all these events, the tribes had no other choice but to unite their forces to put up a defence. The Governor, having left his soldiers in the midst of the tribesmen in the company of Tahir Bey took flight and boarded a vessel which went away. This unaccountable action costs a great number of lives. Amongst those who fell in the field along with the prisoners who had been kept on board was my own brother. It is a plain fact that such evil actions perpetrated against its loyal subjects shall not be connived at by the state. Under the circumstances, I humbly beg that an official be sent here to investigate the cruelties and oppression to which the people were subjected. The entirety of Qatar's population are loyal and obedient subjects of the Ottoman State. Former governors and mutasarrifs may be consulted on this issue. The official to be appointed for this purpose will reveal our innocence. We implore your government not to incriminate us in any way whatsoever in maters of justice, righteousness and moderation. We are under the protection of the state. It was Hafiz Pasha who compelled us to commit things we did not

wish to happen and to retaliate. God forbid, if we fail in our duties towards the state of which we are obedient servants. To cut a long story short, the inspector who will investigate the matter will reveal the truth of our statement. Should we prove to be in the wrong at the conclusion of the said investigations, we are ready to suffer any punishment to be meted out.[59]

In short, the Shaikh wanted a fact-finding mission from Constantinople to reveal the truth about the atrocities allegedly committed by the Ottoman forces in Qatar under the direction of the Wali. This letter was written when the Wali was still in Qatar; it focused on his cruelty towards the Qataris. Shaikh Jasim wanted justice from the High Ottoman authorities.

Shaikh Jasim's case was personally taken up by Said Effendi, the Naqib al-Ashraf, leader of the descendants of the Prophet Mohammad (peace be upon Him) at Basra. Said Effendi met Sultan Abdul Hamid II, in April 1893, after meeting Shaikh Jasim and other leading tribal chiefs in Qatar. He proposed to the Sultan a reconciliation between the two parties. Sultan Abdul Hamid II gave no immediate reply to Said Effendi's proposal until he had heard the views of the Ottoman War Ministry. On 12 April 1893, the War Ministry opined that the visit of Said Effendi to Qatar might cause the British, as well as the anti-Ottoman tribesmen in Qatar, to think that the Ottoman government was incapable of taking any military action in Qatar. As the despatch of troops would undoubtedly demonstrate Ottoman military strength, Said Effendi should go to Qatar with an army and accomplish his task backed by the troops. The views of the War Ministry were largely accepted by the Sultan on 12 April 1893.[60]

Meanwhile, on 13 April 1893 on the Sultan's orders, the Grand Vizier submitted a detailed report on Mehmed Hafiz Pasha's abortive Qatar expedition. The Grand Vizier stated that Mehmed Hafiz Pasha went to Nejd and Qatar to suppress any possible anti-Ottoman movement by the Beni Hajir and al-Manasir tribes. However, the Sultan declined to accept the Grand Vizier's explanation and maintained that his report on Qatar differed significantly from other information, which the Sultan had received. On 15 April 1893, the Sultan wrote back to the Grand

Vizier arguing that Mehmed Hafiz Pasha's dictatorial policy was at least partly responsible for the uprising in Qatar:

> From a study of documents brought to our attention concerning the rebellious acts that Jasim al-Thani, the quaim-makam of Qatar subordinated to Nejd dared to commit, we have observed great divergences between those reported by the governor of Basrah and those reported to Said Effendi, a descendant of the Prophet, by the Sheikh of Qatar. The governor, instead of resorting to suitable remedies which would have been more in line with the circumstances had he paid due attention to the given characteristics of Qatar, seems to have preferred to act audaciously in the belief that he was faced with a situation such as in the case of the other kazas, thus intimidating the tribes. Whereas Jasim should have come over to speak with the governor appointed by the Sultan during his visit to Qatar, his withdrawal from localities and towns as if the troops of a foreign nation were marching on him based on false pretexts created an atmosphere of insecurity which rendered an armed conflict inevitable. Even though one assumes that the governor has failed in certain respects, the chastisement of these insolent renegades is only too natural as the disgrace brought upon the honour of our army by this bloody encounter should be mended. Jasim had expressed his wish to be appointed qaim-makam of Qatar upon its capture in 1871 and has not since then shown any mutinous behaviour. The British had claims of protectorate over the Oman tribes in the environs of Qatar, the Muscat Sultanate and the Bahrain Sheikhdom. Their intention to extend the scope of these aspirations as far as Qatar should not be a far-fetched conjecture. Under the circumstances in chastising the instigators of those incidents which took place in Qatar, a more judicious and moderate action would have been more pertinent. Consequently, instead of despatching troops to punish them as the governor of Basrah seems to suggest, it behoves us to clarify

the matter in the first place and unravel the motives for the dispute. Political and military investigations must be conducted in the region, and according to the result obtained, the people and the prominent personalities there must be enlightened about the evil consequences here and hereafter of acts perpetrated against the state, and the Caliphate it represents.[61]

Therefore, Sultan Abdul Hamid II's opinion on the Qatar uprising was neutral and diplomatic. Mehmed Hafiz Pasha could have managed the situation through diplomacy and wisdom. The Sultan was anxious that Mehmed Hafiz Pasha be removed from his post at Basra immediately, as he was solely responsible for the debacle suffered by the Ottomans soldiers in Qatar. The Sultan recommended that a committee of investigation should be formed and sent to al-Bida at once to enquire into the matter and make necessary recommendations. The investigation committee was composed of Colonel Rasim Bey, the Nakib al-Ashraf Said Effendi and Mohammed al-Sabah, Qaim-maqam of Kuwait, or Mobarak al-Sabah.[62] Military options were thus rejected in favour of political and diplomatic efforts.

BRITISH INVOLVEMENT
In this context, it is important to mention that after the battle of Wajbah, the British government unsuccessfully attempted to get involved in the Qatar crisis. When London learned of the uprising, Lord Kimberley, the Secretary of State for India, was convinced that Colonel Adelbert Cecil Talbot, the Resident in the Gulf (September 1891 to May 1893), should be sent to al-Bida to bring about a peaceful settlement of the dispute between Shaikh Jasim and the Ottomans.[63] As there was general agreement between Lord Kimberley and Lord Roseberry, the Secretary of State for Foreign Affairs, Resident Talbot arrived off the coast of al-Bida on board HMS *Brisk* on the evening of 25 April 1893. Talbot's mission to al-Bida was without the sanction of the Ottoman government. The British government felt this to be unnecessary, as it had never recognised the Ottoman government's claims of jurisdiction over the Qatar coast, where it exercised no real authority. On the morning of 26 April 1893, the *Brisk* sailed into the port of al-Bida and anchored near the Ottoman vessel *Merrikh*, about 3 miles from the shore. Since the Wali of Basra was on board the *Merrikh*, Talbot

called on him, accompanied by Captains Streeton and Godfrey. The Resident explained to the Wali the purpose of his mission, which was to investigate the difficulties with Shaikh Jasim and resolve them peacefully. The Wali replied that he was unable to co-operate with him, as he had received no instruction from his government to do so. Mehmed Hafiz Pasha, expressing his grievances against Shaikh Jasim, reiterated that there 'could be no way of settling the question without extirpating the rebel Shaikh Jasim and his confederates, the Beni Hajir and Monasir'.[64] Talbot insisted that in the light of the friendship between the two governments the Wali should cooperate with him to work out a formula for a peaceful settlement of the Qatar-Ottoman disputes. He informed the Wali that his proposed mediation adhered to the terms of the Anglo-Qatari Treaty of 1868. The Wali however declined to accept the Resident's arguments, maintaining that this was an internal Qatari matter and the British should therefore not get involved.[65]

Failing to land on the shore of al-Bida and bring the Wali round to his line of thinking, on 1 May 1893 the Resident proceeded to Wakra on board the *Lawrence*, which had arrived from Bushire to relieve the *Brisk*. Upon arrival at Wakra, Talbot communicated directly with Shaikh Jasim, who was at that time at Zuaiin, about 15 miles inland north of al-Bida. On the afternoon of the following day, Talbot, accompanied by Godfrey and Streeton, met with Shaikh Jasim in one of the mud forts at Wakra and held a discussion. Shaikh Ahmed and Mohammad bin Abdul Wahab were also present at the meeting. After explaining the purpose of his visit the Resident asked Shaikh Jasim to give him an account of the Wajbah battle. Giving a detailed account of the fighting between his people and the Ottomans, the Shaikh explained that the Arabs at al-Bida town had suffered considerable losses as a result of bombardment by the *Merrikh*.[66] The Qatari Shaikh completed his narrative about the battle of Wajbah; at Talbot's request, he spoke of his desires for the future. He wished to retire to a place of safety and lead a peaceful life under British protection anywhere in Qatar except at al-Bida or Zubara. He wanted to transfer the administration of tribal affairs to Shaikh Ahmed, who would be based at al-Bida, so that no tribe could occupy al-Bida under the Ottoman umbrella. The Resident stated categorically that he had no authority to offer Shaikh Jasim protection. However, he would present his pleas to the British government.[67]

On the afternoon of the following day, 2 May 1893, Talbot met Shaikh Jasim again at the latter's request. Shaikh Jasim, reiterating the points made during the previous day's interview, informed the Resident that if he did not receive justice from the Porte, he would teach the Ottoman a lesson by attacking al-Hasa and Qatif; a grand coalition of all the tribes would be formed to carry out this objective. Shaikh Ahmed, who was also present at the meeting, wanted an immediate settlement of Qatar's dispute with the Ottomans. The political crisis was severely affecting the country's pearl trade, as Shaikh Jasim, a leading pearl merchant, was no longer willing to pay the pearl divers in advance because of the uncertainty at home and the Ottoman deployment of a war vessel off the coast of Qatar. Therefore, Shaikh Ahmed, having been nominated by Shaikh Jasim as 'Plenipotentiary Extraordinary', assured Talbot that he would agree to any British decision with regard to the Ottoman-Qatar disputes. Shaikh Ahmed appealed to be given a place of refuge on the Qatar coast:

> As to the document I have received about your arrival to repair the mischief brought about by the Wali Hafiz Pasha and to mediate towards that end, if your mediation is just and equitable we shall be grateful and will agree to it. If we become settled and secure in our native country at any of the Katr [Qatar] ports of which we may approve, excluding El Bidaa and El Zobarah, then, when that has been obtained for us, I bind myself to you to maintain the peace of the sea at the place we live in, in conformity with the former agreement of our late father Muhammad bin Thani with you, and in accordance with the agreements of the friendly Shaikhs (Trucial Chiefs) with the British Imperial Government.[68]

In short, Shaikh Ahmed was sought British protection through implementation of the dormant Anglo - Qatari Treaty of 1868.

Upon his return to Bushire the Resident, in a telegram to Lord Curzon, the Viceroy of India, reported the results of his mission. He urged that Britain should intervene on behalf of Qatar to bring about a peaceful settlement of the disputes. Shaikh Ahmed was ready to pay an indemnity of India Rs. 20,000 to cover Ottoman losses, if demanded, although he was the injured party.[69] Talbot stated that:

El Bidaa is at present deserted by Arabs, and will remain so unless their safety is assured. It would be thus useless to attempt [to treat with] the Turks, and I strongly recommend effort being made to procure their withdrawal and reinstatement independently of Katr Chiefs on footing of Trucial Chiefs, even though the Turks should condone Jasim's action for the present, in order to avert our mediation. I believe they would be willing to pay small annual instalment, not exceeding kerans 9,000 to clear off any reasonable lump sum demanded by Turkey for cession of supposed rights at El Bidaa and withdrawal from Katr.[70]

The Resident was suggesting a British initiative to bring about an Ottoman withdrawal from al-Bida and that the 1868 Treaty be brought back into operation. The India Office referred the matter to the Foreign Office on 12 May 1893, citing Lord Kimberley's willingness to enter into negotiations with Sultan Abdul Hamid II for the withdrawal of he Ottoman troops from al-Bida on the basis of the offers made by Shaikh Jasim and Shaikh Ahmed. If Lord Roseberry, the Foreign Secretary agreed to Talbot's recommendation, Kimberley 'would be willing to undertake that the Indian Government would guarantee the punctual payments of the instalments of the indemnity that might be agreed upon'.[71] Accordingly, Sir Clare Ford, British Ambassador to Constantinople approached the Ottoman Foreign Ministry at Roseberry's instruction. In fact, Roseberry aimed to detach Qatar peninsula from the Ottoman influence and secure the latter's recognition of the independence of Qatar. However, the Ottoman government rejected the British offer of mediation outright and the Ambassador was informed that Sultan Abdul Hamid II, himself had launched an initiative to bring about a peaceful solution of Shaikh Jasim's disputes with the Ottoman local authorities, appointing an investigation commission to report to the Sultan directly, concerning the causes of the Qatar uprising.[72]

Meanwhile, in accordance with the Sultan's order, the Wali of Basra was dismissed and the commission of inquiry headed by Said Effendi concluded its visit to Qatar. The commission reported to the Sultan in June 1893, that Shaikh Jasim had agreed to continue the payment of taxes that were levied on Qatar by the Ottoman government, return the weapons captured from the Ottoman forces during the battle and maintain good relations with the Ottomans.

In return for these concessions, the Ottomans were to allow those residents of al-Bida, who had abandoned their houses during the fighting to return safely to their homes. Furthermore, the commission recommended re-establishing good relations with Shaikh Jasim and accepting his resignation from the Qaim-maqamship of Qatar in favour of his brother Shaikh Ahmed.[73] Although the Porte did not accept Shaikh Jasim's resignation, the latter transferred official responsibilities to Shaikh Ahmed to avoid personal contact with the Ottoman officials in al-Bida and moved to al-Wusail.[74] However, Major James Hayes Sadler, Officiating Resident in the Gulf (June 1893 to January 1894), believed that the Qatari Shaikh's transfer of responsibility to Shaikh Ahmed was more political and tactical than real. Shaikh Jasim did not do so in order to relinquish authority over his tribe, 'though he probably finds in convenient for the present to interpose his brother as a buffer between himself and the Turks'.(75)

CONCLUSION

The occupation of al-Bida, achieved without opposition, was a major achievement, which furthered the Ottoman objective of extending their sphere of influence up to the eastern coast of Qatar. Although Shaikh Jasim, who emerged as the most dominant figure in Qatari politics, initially welcomed the Ottoman landing, his relations with the occupying forces eventually turned sour. Shaikh Jasim clashed with the Ottomans over their attempt to establish a customs house at al-Bida and the proposed establishment of administrative units at Zubara, Wakra and Khor al-Odaid. The purpose of these administrative reforms was to increase revenues for the Ottoman treasury and to assert closer Ottoman control over Qatar. The hostility eventually led to the battle of Wajbah, in which the Ottomans suffered a crushing defeat at the hands of Shaikh Jasim. This was the first major uprising against Ottoman rule in the region. Shaikh Jasim won the battle because of his applying a well-planned military strategy. He not only cut the water supply to the garrison and blocked the channels of communication with the Ottoman provincial headquarters at al-Hasa, but also seized messengers and correspondence, depriving the besieged soldiers of reinforcements. Following the battle of Wajbah, Shaikh Jasim emerged as an important figure in the Gulf and as a force to be reckoned with. The British government, which had been watching the situation in Qatar, despatched a peace envoy, Talbot, to the Qatar coast to mediate a settlement between Shaikh Jasim and the

Ottomans. The Ottomans however, rejecting the British peace mission, themselves came to terms with Shaikh Jasim. Although Shaikh Jasim voluntarily retired to al-Wusail, transferring responsibilities to his brother Shaikh Ahmed, he remained a vital factor in Qatari politics.

Notes.

1. For more details on the Ottoman Expedition to Nejd, see Abstract of the Contents of a despatch, Under-Secretary to the government of India, no. 54, 31 August 1871, L/P&S/5/261; see also Kursun, op. cit., pp. 51-54.

2. Smith, to Pelly, no. 25, 20 July 1871, L/P&S/5/268.

3. Lieut. Colonel Herbert to Medhat Pasha, no. 44, 18 July 1871, L/P&S/5/268.

4. See the statement of Shaikh Mohammad bin Thani, chief of Qatar in conversation with Mirza Abul Qasim, unnumbered, 19 July 1871, L/P&S/5/268; see also Pelly to the Political Secretary, Bombay, unnumbered, 18 July 1871, L/P&S/5/268; see also Appendix no. 3.

5. Pelly to the SGB, no. 847-218, 31 July 1871, L/P&S/5/268; see also Political Secretary, government of Bombay to the Foreign Secretary, Simla, unnumbered, 18 July 1871, L/P&S/5/267.

6. Lorimer, op. cit., Historical, vol. IB, pp. 802-803.

7. As quoted in Kursun, op. cit., pp. 60-61.

8. Kursun, op. cit., pp. 61-62.

9. Lorimer, op. cit., Historical, vol. IB, p. 803.

10. Kursun, op. cit., p. 62.

11. J. A. Saldanha, op. cit., 'Précis of Bahrain Affairs, 1854-1904', vol. IV, p. 35.

12. Ibid.

13. Ibid.

14. Ibid., p. 36.

15. J. A. Saldanha, op. cit., 'Précis of Katar Affairs, 1873-1904', vol. IV, p. 5.

16. Officiating Under Secretary to the government of India (Fort William), to the Resident, no. 2723P, 10 December 1874, R/15/2/29.

17. Edward Charles Ross to Shaikh Isa bin Ali, no. 564, 12 December 1874, R/15/1/203.

18. See Shaikh Isa to Ross, unnumbered, 17 December 1874; Ross to Isa, no. 60, 22 February 1875; Isa to Ross, unnumbered, 4 March 1875; see also Ross to Isa, no. 95, 12 March 1875; Isa to Ross, unnumbered, 23 March 1875; also see Ross to Isa, no. 114, 31 March 1875, R/15/1/203.

19. Ross to Isa, no. 203, 31 May 1875, R/15/1/203.

20. Isa to Ross, unnumbered, 14 June 1875, R/15/1/203.

21. J. A. Saldanha, op. cit., 'Précis of Katar Affairs, 1873-1904', vol. IV, p. 9.

22. Ibid., p. 13.

23. Idris Bostan, 'The 1893 uprising in Qatar and Shaikh al-Sani's letter to Abdul Hamid II', in Studies on Turkish – Arab Relations, no. 2, Istanbul: Foundation for Studies on Turkish – Arab Relations, 1987, p. 83; see also Cox, Officiating Political Resident in the Gulf to Fraser, SGI/FD, no. 332, 16 July 1905, R/15/2/26.

24. Lorimer, op. cit., Historical, vol. IB, pp. 804-805.

25.   Ibid., p. 820.

26.   Ibid., p. 805.

27.   Copy of a journal received from the Senior Commander of the Ottoman Warship stationed at Qatar, BOA, SD 2158/10, Lef. 20, 25 February 1886..

28.   Ibid.

29.   Copy of a letter from the Mutasarrif of Nejd to the Wali of Basra, BOA, SD 2158/1-, Lef. 13, 8 March 1886.

30.   Kurson, op. cit., p. 79.
31.   Ibid.

32.   Translation of a letter from Shaikh Jasim to the Resident, unnumbered, 24 May 1887, FO 78/5108.

33.   Ross to the SGI/FD, no. 150, 16 July 1887, FO 78/5108.

34.   As quoted in Kursun, op. cit., pp. 80-81.

35.   Ibid.

36.   Ibid.

37.   Ibid.

38.   Idris Bostan, op. cit., p. 83.

39.   For more details see Nafiz Pasha to the Secretary General of the Court, BOA, Y. MTV 31/31, Lef. 2, 10 March, 1888.

40.   Lorimer, op. cit., Historical Vol. IB, pp. 821-822.

41.   J. A. Saldanha, op. cit., Précis of Qatar Affairs, 1873-1904, p.34.

42.   As quoted in Kursun, op. cit., pp. 86-87.

43.   Ibid.

44.   H. Walpole (IO) to the US/S, FO, unnumbered, 2 October 1890, FO 78/5108.

45.   Sir William White, *Note Verbale* to the Sublime Porte, no. 128, 9 December 1890, FO 78/5108.

46.   FO to White, no. 67, 14 March 1891, FO 78/5108; for full text see White to Said Pasha, no. 28, 22 March 1891, FO 78/5108.

47.   See Sublime Porte to the British Ambassador, *Note Verbale*, no. 55, 4 July 1891, FO 78/5108.

48.   See *Note Verbale*, British Embassy to the Sublime Porte, no. 76, 26 August 1891, FO 78/5108; for full Text of the Maritime Truce of 1853, see Aitchison, ibid., pp. 252-253.

49.   *Note Verbale*, Sublime Porte to the British Embassy, no. 13, 26 January 1892, FO 78/5109.

50.   S/S for India to the V in C, unnumbered, 11 February 1892, FO 78/5109.

51.   Kursun, op. cit., p. 91.

52.   Idris Bostan, op. cit., p. 83.

53.   Frederick F. Anscombe, The Creation of Kuwait, Saudi Arabia and Qatar, New York: Columbia University Press, 1997, p. 87.

54.   Kursun, op. cit., p. 93.

55.   Anscombe, op. cit., p. 88.

56.   Kursun, op. cit., p. 93.

57.   Residency Agent Bahrain to the Resident, no. 49, 27 March 1893, FO 78/5109; see also Anscombe, op. cit., p. 88.

58.   Lorimer, op. cit., Historical vol. IA, p. 823; see also Anscombe, op. cit., p. 88.

59.   As quoted in Kursun, op. cit., p. 95.

60.   See Minutes by the Ottoman Minister of War, BOA, 4.MTV.76/133, 12 April 1893.

61.   As quoted in Kursun, op. cit., pp. 97-98.

62.   Ibid., p. 98.

63.   Walpole to the US/S, Foreign Department, unnumbered, 20 April 1893, FO 78/5109.

64.   Talbot to the SGI/FD, no. 75, 5 May 1893, FO 78/5109.

65.   Ibid.

66.   Talbot to the SGI/FD, no. 76, 7 May 1893, FO 78/5109; for full Text of Shaikh Jasim's statement on the battle of Wajbah, see Appendix no. 4.

67.   Ibid.

68.   Shaikh Ahmed bin Thani to Talbot, unnumbered, 3 May 1893, in enclosure to Talbot to the SGI/FD, no. 76, 7 May 1893, FO 78/5109.

69.   Viceroy to the S/S for India, unnumbered, 9 May 1903, FO 78/5109.

70.   Ibid.

71.   India Office to the US/S for FO, unnumbered, 12 May 1893, FO 78/5109; see also India Office to the US/S for FO, unnumbered, 14 June 1893, ibid.

72.   Kursun, op. cit., pp. 99-100; see also Ravinder Kumar, India and the Persian Gulf Region 1858-1907: A Study in Imperial British Policy, Bombay: Asia Publishing House, 1965, pp. 126-127.

73.   Anscombe, op. cit., p. 89.

74.   See the Report of the Investigation Committee, BOA, Y.MTV 79/113, 2 July 1893.

75.   Major S. Hays Sadler, Officiating Political Resident in the Gulf to the SGI/FD, no. 130, 25 June 1893, FO 78/5109.

Plate 9: Sultan Abdul Hamid II (1842-1919)
Reigned 1876-1909

CHAPTER SIX

The 1895 Zubara Tragedy and the Aftermath

Zubara felt victim of Britain acting on Bahrain's behalf. The al-Bin Ali's departure from Bahrain under the leadership of Sultan bin Mohammad bin Salamah and their attempt to form a settlement at Zubara with the active support of Shaikh Jasim bin Mohammad bin Thani and the Ottomans, led to the Zubara crisis of 1895. The Bahraini chief was worried that anti-Bahraini forces at Zubara posed a serious threat to the security of Bahrain. Zubara had fallen into the hands of pro-Ottoman elements such as the al-Bin Ali tribe, and this was a severe blow to British interests and prestige in the Gulf in general and Bahrain in particular. Before the al-Bin Ali fleet, which was concentrated at Zubara, could head to Bahrain, British war ships bombarded the harbour of Zubara and forced Shaikh Jasim to make terms with Commander Pelly and dismantle the settlement.

GATHERING STORM

As mentioned in the previous chapter the Ottoman government had already appointed Arab Effendi and Abdul Karim Effendi as administrator of Zubara and Khor al-Odaid respectively. Well before Arab Effendi had taken up his post, the situation in Zubara became precarious and a serious Anglo-Ottoman dispute developed. This was due to the migration from Bahrain of Sultan bin Mohammad, grandson of the former ally of bin Tarif and the chief of the al-Bin Ali tribe of Bahrain, along with his 1,500 followers, because of differences with Shaikh Isa over a shooting. In April 1895, it was reported by Colonel Frederick Alexander

Wilson, Political Resident (January 1894 to June 1897), that Sultan bin Mohammad and his party had arrived at Zubara and wanted to establish a settlement there with the support of Shaikh Jasim.[1] However, Shaikh Isa, the Bahraini ruler, believed that the establishment of a settlement at Zubara under Shaikh Jasim's patronage posed a serious threat to the security of Bahrain. When this mater was brought to the attention of Resident Wilson, he took it seriously.[2] On 22 April 1895, expressing his solidarity with Shaikh Isa, the Resident delivered a sharp warning to Shaikh Jasim:

> It has been reported to me that Sultan bin Mohamed Salamah in defiance of the wishes of his Chief, Sheikh Esa bin Ali, intends to form a settlement at Zobarah and that in this matter he has your support and assistance.
>
> You are well aware that such a settlement at Zobarah is in direct opposition to the will of the Government. I have now received the orders of Government in this matter, and I accordingly have to convey to you a strong and very distinct warning that such a settlement will not be permitted. You will therefore clearly understand that in order to retain the favour and friendship of Government, you must at once abandon any such project, which will not be tolerated.
>
> I confidently expect an early and satisfactory reply from you, as this matter will not admit to uncertainty or delay.[3]

Wilson's warning to Shaikh Jasim was one sided; he was influenced by the Bahraini ruler. It should be mentioned here that Britain was indifferent to Zubara and its main concern was the security of Bahrain, which could be threatened from Zubara. On 26 April 1895, Shaikh Jasim responded, complaining that Wilson's warning was contrary to the 'friendship and good will' which he had been maintaining with the British government for more than two decades. The Shaikh added:

> ...as long as I possess that friendship from you – I shall not look for it to others. But if I do not get from you the assistance and cooperation, it is needless for me to involve myself in the task of guaranteeing in any degree whatever the preservation of security on

126

land and at sea. And I shall leave Katr to its owners
[the Ottomans] and save myself from troubles.[4]

Shaikh Jasim's reply was thus bold and independent. He would
leave Qatar rather than bow to British pressure. However, he
expressed willingness to meet Wilson in person to discuss the al-
Bin Ali's settlement at Zubara.[5]

Wilson refused to meet Shaikh Jasim and issued a strongly
worded warning to Sultan bin Mohammad on 25 April 1895.
Wilson warned Sultan bin Mohammad that his plan to form a
settlement at Zubara would not be allowed and that he would be
'routed' by the British government. However, if he agreed to
abandon Zubara, Wilson's assistant, John Calcott Gaskin, would
help him to regain the favour of Shaikh Isa and to 'receive
indulgence from him'.[6] Sultan bin Mohammad however, was in
no hurry to comply with this order. In his reply to Wilson he asked
for permission to remain in Zubara until the end of the pearl
season, on which his people's livelihood was based. Moreover,
leaving Zubara at once would be difficult as half of their boats had
been 'sent to Bahrain for being equipped for the work'.[7]

Sultan bin Mohammad's appeal for delay was, however,
unacceptable to Wilson. He believed that Sultan bin Mohammad
was trying to buy time to consolidate his position at Zubara and
'for the operation of intrigues to secure its permanence'. The
Resident warned that he was authorized to force the al-Bin Ali,
headed by Sultan bin Mohammad, to evacuate Zubara 'by seizing
their boats if necessary'.[8] Wilson wanted to deal with Zubara as
soon as possible when he heard that Ibrahim Fauzi Pasha, the
Mutasarrif of al-Hasa, had sent masons and soldiers to Zubara and
had himself arrived there on 5 May 1895, to further the work on the
settlement. Shaikh Jasim, acting in concert with the Fauzi Pasha,
was preparing to hoist the Ottoman flag at Zubara and mosques
and houses were under construction.[9] The Resident was worried
by reports that on 15 May 1895, Shaikh Jasim had helped Sultan bin
Mohammad equip his boats for pearling while Nasir bin Mubarak,
another Bahraini outlaw, who had been living in a camp between
al-Hasa and Oqair, was expected on the scene. While the
Mutasarrif left Zubara for Qatif via Bahrain, Sultan bin
Mohammad appealed to the Porte for Ottoman support and
protection for himself and his tribe at Zubara.[10]

Both the Resident and the Viceroy of India saw the
increased involvement of the Ottoman local authorities in the
affairs of Zubara as a serious menace to Bahrain.[11] However,

before Wilson took direct military action against the al-Bin Ali settlement at Zubara, the British Ambassador at Constantinople was instructed to warn the Porte that Britain was determined to protect the chief of Bahrain from Ottoman aggression.[12] Accordingly, in June 1895, the Ambassador informed the Ottoman Minister for Foreign Affairs that the al-Bin Ali tribe had reportedly settled at Zubara with the active support of the Ottoman Mutasarrif of al-Hasa, with the aim of invading Bahrain. The Ottoman Foreign Minister denied the allegation, stating that he had 'no knowledge of the question'. The British Ambassador reminded the Foreign Minister that Bahrain was under British protection.[13]

Meanwhile, his new allies, Nasir bin Mubarak and the Beni Hajir tribe strengthened Sultan bin Mohammad's position. Wilson was quick to intervene on behalf of Bahrain. With backing from London, the Resident issued a strong warning requesting Sultan bin Mohammad 'to return immediately with his tribe to Bahrein' and to express allegiance to the chief of Bahrain or face British military action. Captain J. H. Pelly, commander and senior naval officer in the Gulf, accompanied by Gaskin, Assistant Political Resident, proceeded to Zubara on board the *Sphinx* on 7 July 1895, to deliver Wilson's warning to Sultan bin Mohammad personally.[14] Pelly was instructed to impress upon Sultan bin Mohammad the necessity of achieving 'an amicable solution of the existing difficulty'.[15] If diplomacy failed, Pelly should adopt an uncompromising attitude:

> ...You will see that Shaikh Sultan is clearly warned that in the event of his hesitation to comply with the requisition now made upon him, measures to enforce it will no longer be delayed. I have therefore to request that should Shaikh Sultan's reception of the orders and warning now addressed to him not be fully satisfactory, you will, after giving him such time as in your judgment many, in the circumstances, be reasonable, proceed to seize such boats belonging to him and to his tribe Al bin Ali, as you may be in a position to capture, in order to enforce submission to the orders of Government.[16]

Wilson was thus determined to take prompt action against Sultan bin Mohammad and his allies by seizing their means of communication.

PELLY'S ZUBARA PROCEEDINGS

Upon arrival on the harbour of Zubara on the afternoon of 7 July 1895, Pelly asked Sultan bin Mohammad to come on board his vessel and receive Wilson's letter of warning. Sultan bin Mohammad claimed to be unable to do so because of illness. Pelly therefore sent Gaskin, accompanied by Residency agent Mohammad Rahim, to deliver the letter to Sultan bin Mohammad and demanded that he reply to it within six hours.[17] On their way to Sultan bin Mohammad's residence, they were stopped by two Ottoman soldiers on the orders of Syed Mohammad Raouf Effendi, mudir of Oqair and agent of the mudir of Zubara, who asked them to call on the mudir. Gaskin, explaining the purpose of their visit to Zubara, told the soldiers that they saw no need to meet the mudir as his presence in Zubara was not recognised by the British government. After a delay of nearly half an hour and fiery exchanges between the two sides, Gaskin was able, with some difficulty, to reach Sultan bin Mohammad's house and deliver Wilson's letter. The al-Bin Ali chief reacted angrily, stating that:

> ...he was no slave or subject to Shaikh Esa bin Ali, and though he could not contend against the power of the British Government, no force or action on their part would compel him to return to Bahrein; that both he and Shaikh Jasim bin Mahomed bin Thani had a few days previously written to Shaikh Esa bin Ali on this subject, and private negotiations were more likely to lead to their conciliation than threats from the British Government, with whom he has nothing to do.[18]

Sultan bin Mohammad was defiant and told Gaskin categorically that 'if the man of war was to wait for a month', he would not reply to Wilson's unjustified letter.[19]

Gaskin and his team returned to the *Sphinx*. In view of Sultan bin Mohammad's uncompromising attitude the *Sphinx* swung into action at 11-am of 8 July 1895, seizing eight al-Bin Ali boats in the harbour of Zubara. The captured boats were handed over to the chief of Bahrain.[20] Effendi protested to Pelly in writing about the *Sphinx's* actions, maintaining that Zubara was a part of Qatar and was the property of the Ottoman state and that the al-Bin Ali were the inhabitants of Qatar 'from ancient days, from grandfather to grandson'.[21]

Asking Pelly to leave the Zubara harbour at once, the mudir warned:

> I shall refer to my Government and you should refer to yours. You have taken those vessels which were in the harbour, in some of which there are (some) pearls, shells, etc. I do not understand your desires (i. e. actions), and this is contrary to (all) laws and usages; and if you have any reason for the same, negotiate with me and receive my reply and prevent your committing these injustices. If any of them (i. e. boats) disappear and anything lost therefrom, you are answerable.[22]

In fact, Pelly violated the international law and conventions by visiting Zubara with an ultimatum without obtaining prior permission from the Ottoman authorities or Shaikh Jasim. The *Sphinx* left the harbour of Zubara on 8 July 1895, for Bahrain to hand over the al-Bin Ali boats, which he had seized to the Bahraini Shaikh.[23]

Shaikh Jasim, who had been watching the situation very closely, quickly intervened on behalf of Sultan bin Mohammad. On 10 July 1895, he wrote to Wilson complaining against Gaskin's unlawful action in Zubara:

> When I learnt the news of the arrival of the man of war, and knowing that Sultan (bin Mohamed Salamah) was ill, I started with the object of meeting the Captain, the Assistant and Khalid (bin Ali). I had scarcely reached Zobarah when I learnt that they had suddenly landed there and frightened the people of the country by (a demonstration of) armed boats, and that the Assistant had landed with a drawn revolver at Sultan (bin Mohamed Salamah); but they (Sultan and his people would appear to be meant) avoided them by good means.
>
> Afterwards they (armed boats) disturbed the (peace of the) sea and seized the boats belonging to the poor and humble striving to earn their livelihood, together with the unopened oyster shells they had on board and all their property.

The Al Bin Ali tribe did not leave Bahrein, but on account of the humiliation, which they experienced; they have not been rebels.

I stood guarantee for them to Shaikh Esa (assuring him) that no mischief would arise from the side of Zobarah or from its inhabitants, and they have committed no offence. They are poor who earn their livelihood; they are not slaves.

Their original domicile was Katr and they resided in Fareybah [Furaiha] and Zobarah, and thence removed to Hawayla and El Bidaa. They have not been ruled as slaves.

This proceeding (which has been undertaken against them) is contrary to your (sense of) justice, and is a wrong unbecoming of you.

They have no knowledge of their own affairs; these are known to God and then to me.[24]

It appears that Shaikh Jasim had already arrived at Zubara and joined forces with the al-Bin Ali tribe, headed by Sultan bin Mohammad. His presence undoubtedly boosted the morale of Sultan bin Mohammad and his people.

Tension had increased when Shaikh Jasim became involved and Sultan bin Mohammad refused to return to Bahrain. Pelly was instructed to proceed to Zubara once again to watch the situation and seize more of al-Bin Ali's boats if he thought it 'expedient'.[25] Pelly arrived at the harbour of Zubara on board *Lawrence* on the afternoon of 13 July 1895. During his stay at Zubara, Pelly learned that Shaikh Jasim was determined to extend all manner of support to the al-Bin Ali despite anything Britain might do and that Shaikh Jasim had advised the al-Bin Ali pearling fleet to proceed to the banks off al-Bida. He also sent several Bedouin tents and camels to help out the al-Bin Ali tribe. Having gained this information about the gathering at Zubara, Pelly left the harbour on the morning of 14 July 1895, in the *Lawrence* to observe developments along the Qatar coast as far as Ras Lafan.[26] On 15 July 1895, the Lawrence encountered nine al-Bin Ali boats proceeding towards Zubara. Pelly seized them, towed them to Bahrain and handed them over to the chief of Bahrain. Sixteen al-Bin Ali boats had now been seized in total.[27]

OTTOMAN INTERVENTION

The Ottomans retaliated on 21 July 1895. Effendi detained nine Bahraini boats sent to Zubara by the chief of Bahrain, to bring back to Bahrain some leading al-Bin Ali families, who wished to return there. Pelly intervened, stating that Ottoman authority at Zubara was not recognised by the British government and demanding that Effendi should tell him under 'what authority' he had detained Bahraini boats.[28] Defending his actions, Effendi wrote back to Pelly that those families were the subjects of the Ottoman government 'from ancient days and forever, and they only settled at Zobarah by the orders of the Mutasarrif of al-Hasa'.[29] Effendi went a step further:

> Both you and Mr. Gaskin have committed injustices beyond rules and friendship in taking boats belonging to the people of Zobarah, goods being in them to the value of 5 lakhs of dollars, and the owners firstly under the protection of God and secondly under the High Ottoman Government. You have also prevented them from following their occupation. So far you have caused loss to all boats belonging to the people of Katr which have reached your side. Also you have seized the boats which you have found belonging to the people of Zobarah, though they have not sinned; and no one has any claims against them except that they have come under the shadow of the Commander of the Faithful and the protection of the High Ottoman Government.
>
> Let it not be hidden from you that Bahrein is under Katr, also Katr and Bahrein are part of the domains of the High (Ottoman) Government. The original persons who took Bahrein from its people were of the inhabitants of Katr. Al Khalifah were merchants from Koweit and strangers to them. They were only four persons and were appointed Agent over Bahrein; also paid tithes and taxes to Katr and to Bin-Saood, the Sultan's officer. Differences arose between them and Bahrein. Then flew the flag of the High Ottoman Government. You falsely represented (matters) to the High Government in support of your claim of protection over them. This not satisfying you, you have

exercised false aggression towards the people of Katr, taking all their boats and goods, also imprisoning (some of) their people, though they are under the protection of God and that of the Commander of Faithful, the successor to the Prophet of God of the world.[30]

In short, Effendi reminded Pelly of the story of the al-Khalifa's migration to Zubara from Kuwait. Their conquest of Bahrain was made possible by the contribution and sacrifices made by the major tribes of Qatar. As the al-Khalifa paid taxes or tributes to the Qataris particularly during the period of Rahma bin Jaber's ascendancy, Effendi believed that Bahrain belonged to Qatar. He further asserted that both Qatar and Bahrain were the parts of the Ottoman empire, which was not acceptable to Britain.

The Zubara crisis now took a more serious turn. It was reported by Pelly that Ibrahim Fauzi Pasha, the Mutasarrif at al-Hasa, had started collection of levies for war. He also instructed Shaikh Jasim to send reinforcement to Zubara. The Ottoman gunboat *Zuhaf* had already arrived at Qatif from Basra on 21 July 1895. In view of this new development, Pelly sought another man of war to save Bahrain in the event of an attack by al-Bin Ali coalition.[31] *Lawrence* was being sent to strengthen Pelly's position.[32] The situation was further aggravated when on 30 July 1895, the *Zuhaf* arrived at the harbour of Zubara and Shaikh Jasim ordered all al-Bin Ali pearling boats as well as other Qatari boats to return to Zubara. He also directed Nasir bin Mubarak and his other followers to proceed there at once.[33] Qatari boats, flying Ottoman flag, were also stationed off the coast, from Ras Lafan to Fasht al-Dibel, in order to intercept any Bahraini intruders. On 31 July 1895, Pelly once again arrived at Zubara to observe developments and learned that Shaikh Jasim was in the vicinity of Zubara, with his fifty followers armed with Martini-Henry rifles.[34] Ottoman meddling and Shaikh Jasim's active involvement convinced Pelly that an attack on Bahrain was eminent. Pelly therefore requested another man of war and extra troops to protect and preserve British prestige.[35]

At the same time, the British Ambassador to the Porte protested against the actions of Ottoman officials and the concentration of forces at Zubara, making it plain that Britain had never recognised Ottoman jurisdiction on the coast of Qatar. Denouncing the Ottoman claims on Bahrain, the Ambassador made it plain that 'all Turkish claims to Bahrein, which is under the protection of the Queen of England, are totally inadmissible' and that necessary measures would be taken to protect the island from aggression.[36] Despite this British representation to the Porte, the atmosphere at Zubara remained the same and preparation for war continued. On 14 April 1895, it was reported that the Mutasarrif of al-Hasa had recruited a large number of tribes from al-Hasa. Together with some Ottoman troops, these were camped near a well named Beer Jejim, situated outside al-Hasa and were ready for war.[37]

BOMBARDMENT OF ZUBARA

The Zubara crisis reached its climax when Fauzi Pasha issued an ultimatum on 19 August 1895 accusing Wilson 'of breaking the peace of the Nejd coast by seizing boats'. The Mutasarrif made it clear that he would no longer restrain the Qatari people from attacking Bahrain and asked the Resident to evacuate British subjects from Bahrain and return the captured al-Bin Ali boats within seventeen days.[38] While Fauzi Pasha threatened, Shaikh Isa lost his nerve, anticipating that Shaikh Jasim and his allies would strike Bahrain from four sides:

> Jasim and the Katr people were to attack from the direction of Manamah, Sultan Bin Salamah and the Al Bin Ali from near Moharrak, Naser Bin Mubarek and his Beni Hajir near Ras-el-Barr, and the men of the Mutasarif, viz., Abdul Rahman Bin Salamah and Selman Effendi with the Mudir from the West (of Bahrein). This was their main project.[39]

It is difficult to say whether Shaikh Jasim had made plans to attack Bahrain. However, Shaikh Isa's aim was to get the British involved in the crisis and force them to take action against al-Bin Ali concentration at Zubara including their boats.

Shaikh Isa's alarming message to Wilson created excitement at Bushire and London. The Resident, with backing from London, directed Pelly to fire on the hostile flotilla anchored in the harbour of Zubara.[40] Upon his arrival at Bahrain on board *Sphinx*, Pelly despatched *Pigeon*, under the command of Lieutenant Commander Cartwright, to keep a strict watch upon the movements of hostile boats off the coast of Qatar. The *Pigeon* found about 200 boats belonging to Shaikh Jasim anchored off Ras Umm al-Hasa.[41] On 5 September 1895, the *Pigeon* further reported that Effendi, the Ottoman mudir, had been on board the vessel and had ordered it to leave Zubara immediately or face Ottoman attack. He also told Commander Cartwright that Shaikh Jasim was determined to attack Bahrain with the boats assembled on the harbour of Zubara and that the Ottomans were going to join the attack.[42]

Pelly thought that an attack on Bahrain was now imminent. To prevent such an eventuality, Pelly arrived at Zubara on board *Sphinx* on the afternoon of 6 September 1895. Before taking action against the hostile boats, however, Pelly sent a letter of warning to Shaikh Jasim, stating that one hour after receipt of the letter, he would commence operations against the fleet assembled at Zubara to attack Bahrain.[43] True to his word, Pelly made his move at 4.45-pm of 7 September 1895, with help from the *Pigeon*. The bombardment continued until forty-four boats had been destroyed, wrecking the (allegedly) planned invasion of Bahrain. A flag of truce was hoisted on shore and all the Ottoman officials, including Effendi, left the town. The Ottoman flag was nowhere to be seen.[44] It is interesting to observe how one-sided the whole Zubara operation was; neither Shaikh Jasim nor the Ottomans returned fire, despite their thunderous declarations. The Ottoman war vessel *Zuhaf* left Zubara well before the arrival of Pelly's fleet.

However, no sooner Pelly stopped firing of his guns than Shaikh Jasim wrote to him, explaining his helpless position in the whole matter:

> When I came to this place (Zobara) my object was no other than (to promote) good, and from the time I came here I sued for your pardon and (begged you) to abstain from unlawful deeds. I complied with your wishes on that very day.
>
> My former letters to the Resident (Assistant Resident) and Mahomed Raheem are with you.
>
> Again when Sheikh Esa became obstinate, I deputed to him my son Khalifah, who begged for 15 days' respite, which, however, was not obtained. And you are acquainted with all that.
>
> When I did not attain my object, and the Mutasarrif's request to you for (the exercise of) good and the restoration of the plundered property was also not granted, he (the Mutasarrif) ordered me to call all the boats (to Zobara) with what object I did not know: whether to intimidate Sheikh Esa into an acquiescence in good deeds; or for the purpose of removing the people from Zobara. And

thus the boats remained (at Zobara) waiting for his (Mutasarrif's) orders.

Had I entertained aggressive designs and resolved to attack Bahrein; I would not have waited for three months. (on the contrary) I was using my efforts towards a good end.

Now all women and children have been destroyed, some by the guns, and others fled to the desert and perished there, without my knowledge or any intimation from you.

I now sue for your pardon, kindness and assurance of safety (aman). I do not entertain (feelings of) opposition towards you. Give permission to the owners of the remaining boats to go to their respective places, and salaams [greetings].[(45)]

The above letter made it clear that Shaikh Jasim denied any intention of invading Bahrain. He was in Zubara with his boats under the orders of the Mutasarrif, who did not explain to him the purpose of such mobilization.

Commander Pelly at once responded to Shaikh Jasim's call for peace, giving reasons for destroying his boats:

The Mutasarrif has informed me that the Katr people are determined to attack Bahrein and that he is unable to restrain them. It is evident therefore that you are the cause of this gathering, but as you have asked forgiveness and peace, this is only obtainable under the following conditions:-

(1) The evacuation of Zobara by the Al bin Ali and the return of that tribe to their allegiance to their Chief (Sheikh Esa bin Ali of Bahrein).
(2) The dispersal of all Bedouins now collected at or near Zobara.
(3) The return of the nine boats that you took belonging to the people of Bahrein.
(4) The delivery of all the boats now at Zobara as part security of any indemnity the Political Resident may levy upon you and that these boats be brought by unarmed men and anchored off the ship.

(5) The delivery of the pirate Hamad bin Selman. If no acceptance of these terms is received from you by 4 PM today, I shall proceed on the further destruction of these boats.[46]

Pelly's decision to bombard Zubara and destroy the Qatari boats and those of the al-Bin Ali, was therefore as a result of receiving an apparently false statement from Effendi.[47] Seemingly the Bahraini chief also incited the British to take action against Shaikh Jasim and the innocent al-Bin Ali tribe.

Shaikh Jasim had no alternative but to accept Pelly's harsh terms for peace. However, he was unable to accept Pelly's last demand, that he should hand over Hamad bin Salman:

I accept your demands with the exception of Hamad bin Selman who is our enemy. He has taken a boat belonging to us and our men. By God he is not amongst us and we do not know his whereabouts today.

May it please God in the long run he will reach you.

The remainder of the Al bin Ali tribe and the boats when floated will be sent to you. I do not entertain opposition to your orders first or last and peace.[48]

Pelly accepted Shaikh Jasim's offer, but only if certain conditions were met:

I have received your letter agreeing with my terms. No boat is to move until after day-light tomorrow morning and then they are to be brought out and anchored near the ship.

If I see any of my terms not complied with or that any boat attempts to leave during the night, I shall open fire without any warning.

I trust that you are sincere in this matter and you will not be the cause of bringing more trouble upon yourself and [your] people.

I shall be glad if you will send some intelligent person having your authority for the purpose of explaining matters.[49]

137

Pelly acknowledged Shaikh Jasim's revised offer with suspicion and caution.

POST AGGRESSION SETTLEMENT

On 9 September 1895, Gaskin, at Shaikh Jasim's request, came to Zubara to confirm that the agreed terms were being implemented. Assuring Gaskin of his intention to cooperate fully, Shaikh Jasim disclaimed all hostile intent on his part and claimed that he was ordered by the Mutasarrif of al-Hasa to requisition all the Qatari boats to enable a simultaneous attack on Bahrain by the Bedouins from al-Hasa and Oqair. It is difficult to say whether Gaskin believed Shaikh Jasim's statement on the Ottoman intentions. This contradicted his earlier statement that he did not know it. Effendi had ordered him to bring his boats to Zubara. However, Gaskin was satisfied with the progress of the evacuation of Zubara under the supervision of Shaikh Jasim. It was evident that Pelly's demand for the mass surrender of the native craft in the harbour of Zubara had been met. Seventy-six al-Bin Ali boats had already reached Bahrain and forty more were on the way with the families of al-Bin Ali. Gaskin estimated that about one third of the total numbers of al-Bin Ali were returning. The rest would be evacuated to Bahrain as soon as extra transport was available.[50] Eventually, by the end of September 1895, all the al-Bin Ali had returned to Bahrain and resumed their normal work there; the chief of Bahrain had assured them that they would enjoy the same conditions that they had previously enjoyed in Bahrain.[51] In October 1895, Sultan bin Mohammad, instead of returning to Bahrain, set off for Basra because of his lack of trust in Shaikh Isa. His boat had to anchor at Ras Tanura due to bad weather. Unfortunately, he was killed on board by some Amamarah of Bahrain.[52]

Gaskin, meanwhile, on 21 September 1895, accompanied by Lieutenant Taylor, commander of the *Plassy* once again arrived at Zubara to record Shaikh Jasim's statement on Zubara affairs. On the following day, Gaskin summoned Shaikh Jasim, who was living in a village about 7 miles inland. Gaskin noted Shaikh Jasim's testimony:

> Jasim again disclaimed all hostile intent on his own part, and asserted that the Mutasarif of Hasa had encouraged the Al bin Ali to settle at Zubara, with a promise of protection, had assured him, after the first seizure of Al bin Ali boats, that the Turkish

138

Government would procure their release, directing him to write to Sheikh Esa requiring their restoration, failing which he was to attack Bahrain, and that when hopes of their recovering the boats failed, the Mutasarif ordered him to collect the Katr boats and such men as he could muster for the attack, and sent some of the Bedouin tribes to support him. Jasim begged, for pardon, and for the restoration of the boats taken after the operations of the 6th instant, on the ground that they had been collected under the Mutasarif's orders; he expressed his earnest desire to come under British protection, undertaking to guard the Katr peninsula provided only that he should be defended from attack by sea.[53]

Shaikh Jasim reiterated his earlier statement. Both he and the al-Bin Ali tribe seemingly became the victims of the Ottoman wrong hope and plans. However, in order to drive out the Ottomans, the Shaikh raised the question of British protection of Qatar, for the second time. While Gaskin made no comment on Shaikh Jasim's plea for protection, he declined to accept the Shaikh's version of the Zubara crisis:

I am entirely unable to accept Jasim's disclaimers, and consider his responsibility is fully established, whatever may have been the motives actuating himself and the Turks in their mutual relations in these proceedings. It has been said that Jasim hoped for a loss of Turkish prestige and authority as the result of the course adopted, which would enable him to recover a practical independence; it has also been rumoured even at Bushire, that the Turks secretly desired that Jasim, whose effective subjection they have been unable to compass, and who inflicted a most disastrous reverse on their troops in 1893, should be crushed by being driven into a collision with British power.[54]

It seems possible, therefore, that the Ottomans helped to bring about the crisis at Zubara, using the British to destroy Shaikh Jasim for his successful anti-Ottoman uprising in 1893. Shaikh Jasim also wished to see the end of the Ottoman occupation of al-Bida. Shaikh

Jasim delivered ninety boats to Gaskin to be taken to Bahrain as he had agreed. He was asked to send the remaining fifty boats, after making necessary repairs and removing the huts built at Zubara by the al-Bin Ali, within fourteen days, which the Shaikh agreed to do without hesitation.[55] However, in the name of humanity, Shaikh Jasim appealed for the return of the seized and un-destroyed boats to their owners so that they could earn their livelihood; those boats were used for pearl fishing and were their main source of income.[56]

Having dealt with Zubara, Gaskin submitted a memorandum to Wilson on 27 September 1895, detailing the British naval action in the harbour as well as Shaikh Jasim's pleas for the return of the boat and his desire for British protection. The memorandum urged the British government to reject Shaikh Jasim's request, unless he produced strong evidences showing his innocence in the Zubara disturbances as a whole.[57] Taking the Bahraini side because of British commercial and political interests in Bahrain and blaming Shaikh Jasim for supporting al-Bin Ali tribe against Bahrain, the Resident endorsed Gaskin's memorandum. The Resident suggested collecting an indemnity from Shaikh Jasim before restoring boats. Wilson calculated the valuation of the captured boats:

Boats of Qatar.......	Indian	Rs. 48,100/-
Boats of al Bu Kawara tribe		Rs. 13,170/-
Boats of Sultan bin Mohammad bin Salamah ...		Rs. 4,550/-
	Total	Rs. 65,820/-

It appears that the boats belonging to the al-Bu Kawara tribe and Sultan bin Mohammad were worth Rs. 17,720. The Resident concluded that Shaikh Jasim should pay up to Rs. 50,000/- and upon the payment of this fine, the captured boats could be returned to him, except Sultan bin Mohammad's boats, which should be retained because of his rebellion against the pro-British Shaikh Isa.[58] While the government of India rejected Wilson's unjust indemnity claim, it maintained that Sheikh Jasim could 'scarcely be given British protection' considering local circumstances.[59] Although the governor general in Council agreed with Wilson that Shaikh Jasim had been 'the main instigator' of the disturbances at Zubara, the Council believed that the fine proposed by Wilson was close to the estimated value of the detained boats and unjust. The Council, therefore, maintained that a fine of Rs. 30,000/- would be sufficient. Regarding Sultan bin Mohammad's boats, the Council suggested these should be handed over to Shaikh Isa rather than to the descendants of the late Sultan bin Mohammad, thus offending not only local law and custom but also logic. On 10 January 1896, Wilson was instructed to inform Shaikh Jasim that unless he paid the decided amount within a reasonable time, the seized boats would be destroyed.[60] Wilson asked Shaikh Jasim to pay the ransom money by 17 February 1896 and stated that he himself would come to Wakra to collect it. If he failed to meet British demand, the seized boats would be destroyed. Shaikh Jasim was not the kind of man to bow to British pressure; he declined. The seized boats were torched at Bahrain in May 1896, at Wilson's direction, to the great loss of the Qatari people.[61] It is important to mention that these destroyed boats had been engaged in pearl fishing, and were thus the main source of income for the owners of the boats as well as the divers.

Following the British destruction of the Qatari boats, the Ottomans, who claimed to be the guardians of Qatar, strengthened their naval position around the coast of Qatar by deploying *Zuhaf* to discourage any future British action similar to that at Zubara. At the same time the Ottoman Embassy in London drew Lord Salisbury's attention to the unlawful acts of the Resident in Zubara. The Ottoman diplomatic note of November 1896 stated:

> The Ambassador of Turkey sends his compliments to his Lordship the Marquis of Salisbury and his honoured to communicate to him the following.
>
> Last year an incident occurred at Zubara on the Persian Gulf and the British squadron proceeded to commit hostile acts against said place, destroying a large number of vessels destined for pearl fishing and burning the dwellings of the tribe established there.
>
> As this area has been under the dominion of the Ottoman Empire since time immemorial and the al Bin Ali tribe, having migrated from Bahrein, had taken refuge under the Imperial flag, it is evident that the squadron of Her Majesty the Queen carried out a hostile act quite incompatible with the amicable relations that exist between our two countries.
>
> Although it is some time since the aggression in question took place, the Imperial Government believes it is not too late to make abundantly clear

to the Government of Her Britannic Majesty the illegal nature of these acts, which have done a great wrong to the people of Zubara and placed them in a highly precarious situation, while injuring the prestige of the Ottoman flag, under whose protection the said tribe had placed itself.

The Imperial Government is convinced that its well-founded complaints will be taken seriously.[62]

In other words, the Ottoman note reasserted an Ottoman position that Zubara had belonged to the Ottoman empire from the earliest times and the actions of the British ships in the harbour of Zubara was illegal and unjustified.

Nevertheless, the following month, the British Foreign Office defended Pelly's act and countered the Ottoman charge:

...They [HMG] consider that the measures in question were necessary for the defence of Bahrein which is under the protection of Great Britain, and they cannot admit the contention that the portion of the coast of the Persian Gulf in which Zobara is situated is within the jurisdiction of the Ottoman Empire.[63]

The British reply was brief and denied Ottoman sovereignty in Zubara; British actions in Zubara were intended to save Bahrain from possible Qatari attack.

THE MINOR UPRISING

The Ottoman volte-face in the Zubara crisis was a great lesson for Shaikh Jasim. He now reviewed his entire relationship with the Porte and reversed his stand, in the light of Ottoman inaction and betrayal during Pelly's operation in Zubara, despite their assurances. Shaikh Jasim got an opportunity to pronounce his uneasiness openly in April 1898, when Shaikh Mubarak al-Sabah, the Kuwaiti ruler, led an expedition against the Beni Hajir tribe and captured a large number of cattle with the support of the Ottoman authorities at Basra. The Beni Hajir tribe was under the protection of Shaikh Jasim and it was his responsibility to guard the tribe's interest. Shaikh Jasim also complained to the Ottomans that Shaikh Mubarak 'had taken amongst the spoils from the Beni Hajir, certain property that belonged to him'.[64] The Ottomans however declined to intervene, turning Shaikh Jasim against them. Consequently, in the absence of any Ottoman gun-boat at al-Bida, in September 1898, the Beni Hajir led an uprising against the Ottoman troops with the active support of Shaikh Jasim; a few personnel were killed on both sides. This was reported by Lieutenant Robinson, the Commander of *Sphinx*, who called at al-Bida harbour in November 1898, on his way to Karachi. During his brief stopover, Robinson met privately with Shaikh Ahmed, who represented Shaikh Jasim at al-Bida; Shaikh Ahmed expressed his willingness to 'turn out the Turks', and enter into agreement with the British government.[65] However, the Political Resident and the government of India were able to ignore Shaikh Ahmed's offer to place Qatar under British protection which he failed to put in writing.[66]

The Ottomans were now in the mood for in reconciliation. Following the minor uprising at al-Bida, Syed Rajab Effendi, Naqib of Basra, was sent to Qatar to mediate between Shaikh Jasim and Shaikh Mubarak in November 1898. Shaikh Jasim, however, declined to meet the Naqib, arguing that the Naqib's last attempt at mediation resulted in a renewed attack by Shaikh Mubarak, in which a number of Qatari people were killed and property was looted.[67] In view of Shaikh Jasim's uncompromising attitude, the Ottoman authorities upgraded their military strength at the al-Bida fort, replacing the existing troops with three new battalions in December 1898.[68] In January 1899 the Ottoman corvette *Zuhaf* was sent from Basra to Qatar to patrol the coast.[69]

Increased Ottoman naval manoeuvres on the Arab littoral together with the Zubara incident, strongly underlined the need, now more urgent than ever, for the British authorities in the Gulf to appoint a permanent British official at Bahrain to counterbalance Ottoman influence both in Bahrain and Qatar. Having obtained Lord Curzon's sanction, Gaskin was appointed as the first Political Agent at Bahrain because of his knowledge of the Arabic language and wider experiences in the affairs of the Gulf, particularly those of Bahrain and Qatar. Gaskin took charge of his office at Manama on 10 February 1900. In addition to Bahrain, his responsibilities included Qatar and the seven Trucial states. He was also responsible for the promotion of British trade and influence in that part of the Gulf, a policy which Lord Curzon advocated throughout his time as Viceroy of India.[70]

CONCLUSION

For geographical and historical reasons, in the hands of their enemies, Zubara was like a dagger pointing at the hearts of the al-Khalifa. The concentration of the dissatisfied al-Bin Ali tribes at

Zubara, with the strong backing of Shaikh Jasim and the Ottomans, was the main reason why the British bombarded the harbour of Zubara. Britain attacked in the name of saving Bahrain from al-Bin Ali attack, relying on the wrong information supplied by the Shaikh of Bahrain. The crisis was soon over and peace was restored on British terms. All the members of the al-Bin Ali tribe except their leader Sultan bin Mohammad were evacuated to Bahrain; all the captured boats were destroyed, to the dismay of Shaikh Jasim. The Ottomans did nothing to resist the British despite their resentment of the British presence in Zubara prior to the bombardment. They, in fact, brought about the crisis by giving the British the wrong signal. Shaikh Jasim had no alternative but to accept the peace terms dictated by the British. However, he reviewed his relationship with the Ottomans, staging a minor uprising against the Ottoman garrison at al-Bida. Shaikh Jasim also sought British protection once again to drive out the Ottoman forces from his land. While the Ottomans strengthened their position both on land (al-Bida) and at sea, the British appointed their first Political Agent to Bahrain to keep a close eye not only on Bahrain but on Qatar also.

Note

1.   Wilson to SGI/FD, no. 42, 4 May 1895, R/15/1/314.

2.   Ibid.

3.   Wilson to Shaikh Jasim, no. 75, 22 April 1895, R/15/1/314.

4.   Shaikh Jasim to Wilson, unnumbered, 26 April 1895, R/15/1/314.

5.   Ibid.

6.   Wilson to Sultan bin Mohammad bin Salamah, no. 76, 22 April 1895, R/15/1/314.

7.   Sultan bin Mohammad to Wilson, unnumbered, 25 April 1895, R/15/1/314.

8.   Wilson to the SGI/FD, no. 42, 4 May 1895, R/15/1/314.

9.   Wilson to the SGI/FD, no. 44, 11 May 1895, R/15/1/314.

10.   Wilson to the SGI/FD, no. 46, 18 May 1895, R/15/1/314.

11.   V in C to the S/S for India, no. 1001-E, 21 May 1895, F.O. 78/5110.

12.   US/S, IO to the US/S, FO, unnumbered, 22 May 1895, R/15/1/314; see also the Secretary of State for India to the Viceroy, unnumbered, 8 June 1895, F.O.78/5110.

13.   Sir P. Currie to Kimberley, no. 401, 22 June 1895, F.O. 78/5110.

14.   Wilson to Sultan bin Mohammad, no. 133, 5 July 1895; see also Wilson to Pelly, no. 2.SZ, 5 July 1895; Pelly to Wilson, unnumbered, 9 July 1895, R/15/1/314.

15.   Wilson to Pelly, unnumbered Memo, 5 July 1895, R/15/1/314.

16.   Wilson to Pelly, Commander R. N. H. M. S. *Sphinx*, no. 2-S.Z., 5 July 1895, R/15/1/314.

17.   Pelly to Sultan bin Mohammad, unnumbered, 7 July 1895, R/15/1/314; Sultan bin Mohammad to Agha Mohammad Rahim bin Saffar, unnumbered, 7 July 1895, R/15/1/314; also see Pelly to Sultan bin Mohammad, unnumbered, 7 July 1895, R/15/1/314.

18.   Gaskin to Pelly, unnumbered, 8 July 1895, R/15/1/314.

19.   Ibid.

20.   Pelly to Wilson, unnumbered, 9 July 1895, R/15/1/314.

21.   The mudir of Oqair and Agent to mudir of Zubara to those Officers of the British government, no. 5, 8 July 1895, F.O. 78/5110.

22.   Ibid.

23.   Pelly to Wilson, unnumbered, 9 July 1895, R/15/1/314.

24.   Shaikh Jasim to Wilson, unnumbered, 10 July 1895, R/15/1/314.

25.   Wilson to Pelly, no. 3-S.Z., 10 July 1895, R/15/1/314.

26.   Pelly to Wilson, unnumbered, 14 July 1895, R/15/1/314.

27.   Pelly to Wilson, unnumbered, 17 July 1895, R/15/1/314.

28.   Pelly to Effendi at Zubara, unnumbered, 23 July 1895, R/15/1/314.

29.   Effendi to Commander Pelly, H. M. S. *Sphinx*, unnumbered, 26 July 1895, F.O. 78/5109.

30.   Ibid; see also R/15/1/314.

31.   Pelly to Wilson, unnumbered, 23 July 1895, R/15/1/314; see also Wilson to the FS, Simla, unnumbered, 24 July 1895, R/15/1/314.

32.   Viceroy to the S/S for India, unnumbered, 27 July 1895, F.O. 78/5110; see also the FS, Simla to Wilson, no. 1454-E, 27 July 1895, R/15/1/314.

33.   Pelly to Wilson, unnumbered, 30 July 1895, R/15/1/314.

34.   Pelly to Wilson, unnumbered, 6 August 1895, R/15/1/314.

35.   Pelly to Wilson, unnumbered, 14 August 1895, R/15/1/314.

36.   Philip Currie, British Ambassador at Constantinople to Lord Salisbury, no. 520, 12 August 1895, F.O. 78/5110; see also P. Currie to the FO, London, no. 84, 12 August 1895, F.O. 78/5110.

37.   Pelly to Wilson, unnumbered, 14 August 1895, R/15/1/314.

38.	See the 'Letter of Proceedings', Pelly to Wilson, no. 41, 24 August 1895, F.O. 78/5110; see also the V in C, Simla to the S/S for India, no. 1669-E, 1 September 1895, R/15/1/314.

39.	Shaikh Isa to Wilson, unnumbered, 15 September 1895, R/15/1/314.

40.	Lord Salisbury to Currie, no. 137, 19 August 1895, F. O. 78/5110; see also the FS to Wilson, no. 1608-E, 22 August 1895, R/15/1/314; Wilson to the SGI/FD, no. 89, 24 August 1895, R/15/1/314.

41.	'Letter of Proceedings', Pelly to Wilson, no. 42, 1 September 1895, F.O. 78/5110.

42.	Pelly to Wilson, unnumbered, 7 September 1895, R/15/1/314.

43.	Pelly (at Zubara) to Shaikh Jasim, unnumbered, 6 September 1895, R/15/1/314.

44.	Pelly (from H. M. S. *Sphinx* at Zubara) to Wilson, unnumbered, 7 September 1895, R/15/1/314; for more details on Pelly's military action against Shaikh Jasim's fleet and the reasons for doing so, see Appendix no, 5.

45.	Shaikh Jasim to Pelly, unnumbered, 7 September 1895, R/15/1/314.

46.	Pelly (at Zubara) to Shaikh Jasim, unnumbered, 7 September 1895, R/15/1/314.

47.	See Appendix no. 6.

48.	Shaikh Jasim to Pelly, unnumbered, 7 September 1895, R/15/1/314.

49.	Pelly to Shaikh Jasim, unnumbered, 7 September 1895, R/15/1/314.

50.	Wilson to the SGI/FD, no. 101, 21 September 1895, F.O. 78/5110.

51. Wilson to the SGI/FD, no. 103, 28 September 1895, R/15/1/314.

52. Kursun, op. cit., p. 106.

53. Wilson to the SGI/FD, no. 103, 28 September 1895, R/15/1/314.

54. See ibid.

55. Lieutenant W. F. Taylor to the Political Resident in the Gulf, unnumbered, 23 September 1895, F.O. 78/5110.

56. Shaikh Jasim to Wilson, unnumbered, 1895, F.O. 78/5110.

57. See the 'Memorandum' from Gaskin to Wilson, unnumbered, 27 September 1895, Enclosure to Wilson to the SGI/FD, no. 110, 19 October 1895, F.O. 78/5110.

58. Wilson to the SGI/FD, no. 110, 19 October 1895, F.O. 78/5110.

59. The SGI/FD to Wilson, unnumbered, 12 October 1895, R/15/1/314.

60. W. J. Cuningham, SGI/Fd to Wilson, no. 68-E, 10 January 1896, F.O. 78/5110; see also A. Godley (IO) to the US/S, unnumbered, 15 April 1895, F.O. 78/5110.

61. Kursun, op. cit., p. 107.

62. See the Note (English translation) Ottoman Embassy in London to Lord Salisbury, unnumbered, 6 November 1896, F.O. 78/5110.

63. Foreign Office to the Ottoman Ambassador in London, unnumbered, 5 December 1896, F.O. 78/5110.

64. J. A. Saldanha, Op. Cit., Précis of Qatar Affairs; 1873-1904, vol. IV, p. 45.

65. Ibid., p. 46.

66.   Ibid.; see also Lorimer, op. cit., Historical, vol. IB, p. 829.

67.   British consul at Basra to the consul-general at Baghdad, no. 780, 20 December 1896, F.O. 78/5110.

68.   Acting British consul-general, Baghdad to Sir Nicholas R. O'Conor, British Ambassador at Constantinople, no. 637, 29 December 1896, F.O. 78/5110.

69.   A. C. Wratislaw, Consul at Basra to consul-general, Baghdad, no. 3, 18 January 1899, F.O. 78/5110.

70.   On the appointment of British Political Agent at Bahrain, see the US/S for IO to the US/S for Foreign Affairs, unnumbered, 7 February 1900, F.O. 78/5110; see also FS, Calcutta to the Resident, no. 335, 2 February 1900, F.O. 78/5110; also see Tuson, op. cit., p. 44.

Plate 10: Top: Model of a Dhow
Bottom left: A Baghla   Bottom right: A Battil
These three models were built in 1851 in Bombay

CHAPTER SEVEN

Struggle for British Protection

Qatar intensified its pleas for British protection, aiming to drive the Ottomans from al-Bida fort. Although the Political Resident in the Gulf backed by Lord Curzon, Viceroy of India, extended his unqualified support to the notion of a protection treaty with Qatar on the Trucial model, London declined to do more than to maintain the *status quo*. While no decision was taken on the question of a protection treaty, Britain's relations with the Sublime Porte became precarious when the latter appointed administrative officials to three vital towns: Zubara, Wakra and Khor al-Odaid. The Ottomans' attempts to strengthen their position in Qatar by establishing administrative units had no impact on the internal affairs of the country. Following the death of Shaikh Ahmed, Shaikh Jasim had once again to take full charge of the administration of Qatar.

THE QUESTION OF PROTECTION

At the beginning of the twentieth century the coast and interior of Qatar was riven by conflict, drawing greater British attention to the peninsula. In June 1900, a fleet of the al-Bin Ali tribe and other boats from Qatar attacked five boats belonging to the Amamarah tribe of Bahrain who were engaged in pearl diving operations off the coast of Wakra. The peace at sea was further disturbed in September 1900, when the Beni Hajir tribe, headed by Salmin bin Yetama, clashed with Abdul Hadi bin Mirait off Dhakira on the north of Qatar coast.[1] As the continuous pearl fishing disputes in the Gulf might disturb the regional peace and produce serious consequences, John Calcott Gaskin, the first Political Agent in

Bahrain (February 1900 to October 1904), recommended that al-Bin Ali tribe should be fined and their properties at Bahrain confiscated until the fines were paid. With regard to the second incident, off the coast of al-Dhakira, both Gaskin and Col. Charles Arnold Kemball, acting British Political Resident in the Gulf (April 1900 to April 1904), maintained that as the Ottoman authority in Qatar was nominal, Shaikh Ahmed, whose authority throughout Qatar was 'sufficiently stable in nature' should be made responsible for controlling the Bedouins living in Qatar and that he should warn all boat owners residing on the Qatar coast that they must not to allow their boats to be used by these Bedouins.[2] The government of India concurred, and the Resident was instructed to deal with Shaikh Ahmed directly in the above-mentioned case. Meanwhile, the al-Bin Ali duly paid the fine of Rs. 1,500 imposed upon them for disturbing the peace at sea off Wakra.[3]

The absence of formal relations between the British and the Qatari authorities was key to grasping the question of British protection of Qatar, which came up to the fore again at the beginning of the twentieth century. Although neither the ruler nor anyone else was legally bound to carry out British order or accept British advice, the Qatari authorities complied with British desires. In this connection it is important to note that both Shaikh Jasim and Shaikh Ahmed had already expressed willingness to conclude a British protection treaty similar to the British agreement of March 1892 with the Trucial chiefs of Abu Dhabi, Dubai, Ajman, Sharjah, Ras al-Khaimah, and Umm al-Qawain. This exclusive agreement obliged these Trucial chiefs not to enter into any agreement with any power other than the British and on no account 'cede, sell, mortgage or otherwise give for occupation' any part of their frontiers, other than to the British government.[4] As no action had been taken on the matter of protection, Shaikh Ahmed applied formally to Gaskin in March 1902. In an urgent message he pleaded:

> ...if His Majesty's Government will extend their protection to him and his following, he will be glad to reside at any place in Katr which the Government may think fit, that he will hold himself responsible to keep the seas round Katr free from pirates, and that he will co-operate with the Government and the Chief of Bahrein in any matters which may concern them on the mainland.[5]

Shaikh Ahmed's offer was therefore conditional. He would preserve the peace at sea around his country only in return for British protection of Qatar.

As Gaskin gave no immediate reply to Shaikh Ahmed's request for British protection, the Shaikh became impatient. In another message on 27 March 1902, the Shaikh reiterated that he needed such protection to drive the Ottomans from al-Bida. Gaskin though was convinced that the Shaikh could easily force the garrison to quit al-Bida without resort to arms, as they were entirely at his mercy as regards supplies and water, which he could effectively cut off.[6] Since their arrival in al-Bida in 1871, the Ottoman troops, it is important to remember, had been staying at an old fort at al-Bida. The Ottomans made no attempt to improve or repair the fort. They merely occupied the site in a defensive or even passive manner, and their tenure relied on the presence of a gunboat.[7]

In spite of the unfavourable position of the Ottoman forces in al-Bida, Gaskin was not in favour of encouraging Shaikh Ahmed to move against them unless Britain went to war with the Ottoman empire. While Gaskin was opposed to the expulsion of the Ottomans from al-Bida, he was in favour of granting British protection to Shaikh Ahmed. Strongly recommending that the Shaikh's request be granted, the Assistant Political Agent stated that the Shaikh was already a wealthy chief and was influential with the 'nomad tribes', and that his influence and prestige would rapidly increase if it was known that he enjoyed British protection. Moreover, he was the only chief capable of maintaining order along the Qatar coast and of keeping the Bedouins under control.[8] Gaskin was convinced that a protection treaty would remove any Qatari threat to Bahrain. Putting forward further reasons for his views, the Political Agent stated:

> Bringing Sheikh Ahmed under our protection at Zobara would be an effectual step towards establishing our policy of not recognising the claims of Turkey to Katr, at the same time diminish the number of places on the mainland from which Bahrein can be threatened, and draw a line across Katr to prevent Turkish encroachment towards the northern end of the peninsula. I would strongly urge the careful consideration of the question now that the opportunity offers itself. Turkey is now

turning her attention to strengthening her position in these regions, and she may go so far as to establish herself in a manner from which it may be difficult to dislodge her hereafter.[9]

Gaskin, in fact, was talking about protecting Zubara only, under the leadership of Shaikh Ahmed, not the whole of Qatar. Kemball, endorsing Gaskin's recommendation, urged the government of India to authorise Kemball to enter into an arrangement with Shaikh Ahmed, recognising his independence and promising protection from interference by other powers, on condition that he should take responsibility for the maintenance of order in Qatar and the prevention of 'piracies by sea'. However, Kemball felt that before any definite arrangement was made with Shaikh Ahmed, it was necessary to ascertain more about his position and how the succession to the rulership would develop on Shaikh Jasim's death; Shaikh Jasim was already over 80 and the tribes regarded him as chief, but as far as the Ottomans were concerned he had retired in favour of Shaikh Ahmed.[10]

The government of India was in no hurry to move further unless it saw a full report on the actual position of Shaikh Ahmed in Qatar. Therefore, Kemball was directed not to proceed beyond the enquiry concerning Shaikh Ahmed, and to submit the required report on him.[11] On 27 December 1902, Kemball, after consulting Gaskin reported that Shaikh Jasim bin Thani was the recognised chief of Qatar, and that the Ottomans had conferred on him the title of Qaim-maqam, though he repudiated it and his brother Shaikh Ahmed bin Thani was regarded as his heir.[12] Curzon, who not only headed the administration in India but was also responsible for overseeing the affairs of Gulf as a whole, responded to the acting Resident's report in January 1903, suggesting that:

> ...to prevent Turkish designs, it would be wise to enter into an agreement, but that if His Majesty's Government were not prepared to take this step at once, which would perhaps depend upon the attitude of the Turks, the Sheikh might be told that Government would prefer to wait till he succeeded later on to the Chiefship.[13]

Though he favoured the proposed Qatar treaty, Curzon's suggestion was far from straightforward. He believed that the British government should defer the question of protection until

154

Shaikh Ahmed had taken full responsibility of Qatar upon the death of Shaikh Jasim although, of course, no one knew when Shaikh Jasim would die.

While Anglo-Qatari attention was focused on the question of protection, the Ottoman government decided once more in November 1902, to establish small administrative units at Zubara and Khor al-Odaid. Later reports from Bahrain confirmed this decision, stating that posts would be established not only at Khor al-Odaid and Zubara but also at Wakra. An officer named Yusuf Beg Effendi had been appointed to Wakra and was waiting at al-Hasa for further instruction from the Porte. Officers for the two other places had not yet been appointed. Sir Nicholas O'Conor, British Ambassador to Constantinople, feared that any British protest against such announcements would further jeopardise Anglo-Ottoman relations, in view of anti British Ottoman feelings with regard to Kuwait. However, the Ambassador suggested a British representation to the Porte, though he expected no positive result.[14] Curzon was at one with O'Conor, reiterating that the British government was unable to enter into agreement with Shaikh Ahmed until he succeeded to the chieftaincy.[15]

The Viceroy's recommendation was considered by the India Office and the Foreign Office in London. Lord Lansdowne, the Foreign Secretary stated that the government of India considered it necessary to prevent an Ottoman occupation of Zubara and Wakra as well as Khor al-Odaid. The former two places were situated on the peninsula of Qatar, where the British government had frequently declared that they did not recognise the sovereignty of the Ottoman Sultan. Khor al-Odaid on the other hand, had been for 'the last 30 years held by the Government of India to belong to the Chief of Abu Dhabi, with whom they had maintained direct treaty relations for many years'. In light of these circumstances, George Hamilton, Secretary of State for India, opined that 'should the maintenance of the status quo in El Katar be arranged, for the present at least, no agreement with Shaikh Ahmed bin Thani will be possible'.[16] Hamilton thus dashed Shaikh Ahmed's hopes for British protection.

MUDIRS FOR ZUBARA, WAKRA AND KHOR AL-ODAID

On 3 March 1903, Arabi Effendi, the mudir designate for Zubara, arrived at Bahrain from Baghdad to take up his post. Abdul Karim Effendi, the mudir designate for Khor al-Odaid, was on the way to Bahrain from Basra while Yusuf Beg Effendi was at al-Hasa waiting to proceed to Wakra. Kemball was convinced that Shaikh

Jasim had asked the Porte to establish administrative units in those three places in order to release himself from responsibility; the Ottoman authorities wished to hold him responsible for the disturbed state of the country between Oqair and al-Hasa. However, Kemball believed that the establishment of the Ottoman posts in Zubara and Khor al-Odaid would embroil the Ottomans in disputes with Britain. Any Ottoman move on Zubara would be a threat to the security of Bahrain, as Kemball stated:

> It is, in my opinion, absolutely essential for the security of the Bahrein Islands that Zobara should not be occupied by the Turks. Apart from the fact that the occupation of Zubara would be viewed with the greatest concern by the Chief of Bahrein, who considers the place to be an appanage of his, and whose rights we are bound to maintain, the prestige which the Turks would gain throughout the countryside by the occupation of this place, in opposition to the well-known views of the British Government, would be so great that an attack on Bahrein from Katr could at any time be organized, and the continual presence of a ship of war in Bahrein waters, and perhaps even a military occupation of the islands, would be required.[17]

Therefore, the acting Resident proposed that the British warship *Sphinx* should be sent to Zubara and Khor al-Odaid to prevent the Ottoman officials from landing in those places. Both the Viceroy and the Foreign Secretary approved the acting Resident's proposal. Accordingly, O'Conor was instructed, on 19 March 1903, to make representation to the Ottoman government, making it clear that the British government had never recognised their claim to sovereignty over the Qatar peninsula and that they should maintain the *status quo* in that region. However, O'Conor was not in favour of making such representation to the Porte. The Ambassador maintained:

> The question of Turkish sovereignty over Katar promontory has been repeatedly the subject of discussion with the Porte, and I am not without hope of succeeding in dissuading them from proceeding with the appointments of Sub-Governors, on the ground that any disturbance of

the status quo is impolitic, without having to do more than to remind them of the declarations made by His Majesty's Government on this subject on previous occasions.

I confess I see strong objections to basing our representations on the fact that we do not recognise Turkish sovereignty over the coast, as its immediate result will be to provoke the Ottoman Government to reassert, in the most formal and official manner, their claim, as happened in 1891-92 ... and if we allow their pretensions to pass with nothing more than a verbal or written contradiction, we shall rather have weakened than strengthened our present position. If, on the other hand, we are not satisfied with merely disclaiming their pretended rights, and the Porte persist in their intentions, we shall be forced to take some definite action which may lead His Majesty's Government further than they intended, or further than it is necessary or politic to go at present. At any rate, before taking a step which may raise a very vexed and complicated question, it will, I venture to think obviously be advisable to decide beforehand whether His Majesty's Government will forcibly prevent the Ottoman Government from establishing any new outward symbols of Turkish sovereignty along the whole west coast of the Persian Gulf to Katif, and if so, what will be their policy and action in regard to the Turkish military station at Bidaa, which undoubtedly exists and has existed for many years.

It may safely be said that at the present moment Ottoman sovereignty over the whole of Arabia is, at the best, a very slender one, and that, in the gradual process of disintegration, which is proceeding pretty rapidly, it will become less and less. Is it, under these circumstances, in accordance with British interests to hasten this disruption, or even to over weaken Turkish rule until we have some other to put in its place? It may be that, in the present instance, His Majesty's Government is prepared, after the death of the ruling Chief of Katar, to recognise his brother Ahmed-bin-Thani, and profit by his friendly disposition to conclude with him some agreement on the lines of those with the Trucial Chiefs, which will give us a sort of Protectorate or lien over the promontory; but until this days arrives, it is not clear to me that we gain anything by formally raising without absolute necessary, the question of sovereignty, such as it is. I do, however, think it very possible that, by doing so, and by overtly pursuing a policy which, rightly or wrongly, is considered as directed against the integrity of the Ottoman Empire, we encourage another Power to some act of rank aggression which will precipitate developments, and possibly place us, before very long, in a serious and grave predicament.

There is no doubt that our action in regard to Kuweit has made a very painful impression upon not only the Sultan, but upon many of his Anglophil subjects, whose weight and influence was a counterpoise to the constant pressure and intrigue of that power, whose object it is to persuade both the Sovereign and his ministers that the interest and salvation of Turkey was in a close and cordial understanding with them.

Besides these few remarks which I have made bold to lay before Your Lordship, I beg leave to say that I find some difficulty in clearly understanding the policy proposed by the Government of India in regard to Odeid. On the other hand it is suggested that our warning to the Porte should be based on the fact that we consider Odeid as an appanage of the Chief of Abuthabi, an independent Ruler with whom we have Treaty relations, that is to say, that we consider Odeid as independent of the Chief of Katar, while, on the other hand, we appear to be contemplating the contingency of counteracting Turkish designs in Katar by countenancing the absorption of Odeid by Ahmed-bin-Thani.[18]

In other words, unlike the Foreign Secretary, O'Conor believed it was unwise to object to the Ottoman appointments in Zubara, Wakra and Khor al-Odaid on the basis that the British had never recognised Ottoman sovereignty on the coast of Qatar. Such a move on the part of Britain would simply encourage the Porte to reassert sovereignty not only in Qatar but also Bahrain. The issue of Ottoman sovereignty should therefore be deferred until the death of Shaikh Jasim and the assumption of the full power by Shaikh Ahmed, with whom a protectorate treaty could be signed on the Trucial model, as recommended by O'Conor.

Meanwhile, the Ottomans had taken effective measures to realise their design in Qatar. While an Ottoman gunboat left Basra for Qatar, Yusuf Beg Effendi, mudir designate of Wakra, had already left al-Hasa and arrived at Bahrain on 21 April 1903. Kemball was determined to foil Ottoman plans, pleaded with the government of India to urge the Ottoman authority in Basra to order Yusuf Beg Effendi not to take up his appointment at Wakra. At the same time Gaskin was instructed to detain Yusuf Beg Effendi in Bahrain unless he returned to Basra. However, to the great dismay of Kemball, the Wakra mudir reached Wakra on 27 April 1903, escaping detention.[19] While Yusuf Beg Effendi left Bahrain, the Ottoman Grand Vizir informed O'Conor that he had asked for explanation from the Wali of Basra regarding Yusuf Beg Effendi's appointment. At the same time he assured O'Conor that the *status-quo* would be maintained in Qatar.[20]

Yusuf Beg Effendi, as it turned out, faced hostility from the people of Wakra, who forced him to go back to al-Bida on 30 April 1903; he soon returned, however, and succeeded in establishing himself there through the intervention of Shaikh Ahmed, who feared he would get into trouble with the Ottomans if he refused to assist the mudir. Yusuf Beg Effendi's reinstallation angered the government of India. At the Viceroy's recommendation, the Secretary of State for Foreign Affairs instructed O'Conor to inform the Porte that the British government could not allow Yusuf Beg Effendi to remain at Wakra and the Porte should take immediate step to remove him.[21] The Ambassador was assured that the Ottoman government had no knowledge of Yusuf Beg Effendi's appointment and that he would be recalled. London was dissatisfied with this assurance. Kemball was informed that if he considered Yusuf Beg Effendi's presence at Wakra undesirable, even for a short period, and was of the opinion that a British warship would convince him to leave without raising local difficulties, then this might be arranged.[22]

Kemball was sceptical. He stated that the forceful removal of Yusuf Beg Effendi would inevitably cause local difficulties. Moreover, since Yusuf Beg Effendi had gone to Wakra on the orders of his superior officer, he would refuse to comply with the demands of the commander of a British ship. Therefore, he recommended that the Porte should be asked to issue fresh and definite orders for Yusuf Beg Effendi's immediate recall.[23] While the Anglo-Ottoman row over Yusuf Beg Effendi's removal continued, Arabi Effendi, mudir designate for Zubara, tired of waiting at al-Hasa without salary, returned to Basra in May 1903. Consequently, in June 1903 the Wali of Basra instructed the Mutasarrif of al-Hasa not to appoint more mudirs until further orders. At the same time the mudir designate for Khor al-Odaid, Abdul Karim Effendi, was recalled to Basra from al-Hasa, where he had been waiting to take up his post. Eventually, in June 1903, his appointment was cancelled and no new mudir was appointed to Khor al-Odaid.[24] In July 1903, it was reported that Yusuf Beg Effendi had been dismissed from his post at Wakra, while Shaikh Abd al-Rahman, son of Shaikh Jasim bin Thani, was appointed to replace him on the orders of the Mutasarrif of al-Hasa. As this new appointment at Wakra was contrary to the Porte's earlier assurances to O'Conor, Gaskin felt that further representation should be made to the Porte, urging the Ottoman government to abolish the post and to revert to the *status quo ante*.[25]

While O'Conor made fresh representation to the Porte in regard to the appointment of Shaikh Abdur Rahman at Wakra, stating that it was contrary to the *status quo* in Qatar,[26] Gaskin arrived in Qatar on 16 September 1903, at Kemball's instruction, to obtain a reliable account of Qatar affairs. Upon his arrival in the peninsula, the Assistant Political Agent interviewed both Shaikh Jasim and Shaikh Ahmed. The former had been living in al-Wusail, a coastal village about 12 miles north of al-Bida, with his own household and two or three allied families since the battle of Wajbah. He informed the Political Agent that he had retired from the government of the Qatar peninsula and had transferred responsibility for its future welfare to his brother Shaikh Ahmed. He further stated that he had informed the Ottomans that they should refer all matters to Shaikh Ahmed and then severed all connections with them. The Ottoman authorities at al-Bida had been doing so since 1898. As regards his son, Shaikh Abdur Rahman, Shaikh Jasim stated that he was made Shaikh of Wakra five years ago by Shaikh Ahmed and that though he had heard about the recent actions of the Mutasarrif of al-Hasa he had paid no attention and left Shaikh Ahmed to deal with the matter. Gaskin's conversation with Shaikh Jasim convinced him that the latter disapproved of his son's appointment by the Ottomans as the mudir of Wakra.

Following the conclusion of the al-Wusail meeting, the *Lawrence*, carrying the Political Agent, arrived off Wakra the next day and discussions were held with Shaikh Ahmed. The Shaikh confirmed his brother's statement on the Ottoman appointment of Shaikh Abdur Rahman as mudir of Wakra on salary of 6 Ottoman liras per month, following the withdrawal of Yusuf Beg Effendi. However, the Shaikh stated that Shaikh Abdur Rahman, who had already been Shaikh of Wakra for five years, had not yet drawn the salary offered to him and gave no guarantee for his future loyalty to the Porte despite Ottoman insistence. Shaikh Ahmed took the opportunity to remind the Political Agent of his earlier pleas for British protection, arguing that:

> ...the Mutessarif is apparently anxious to extend the Turkish authority in Katr; and if he succeeds in his designs the whole peninsula will be absorbed by the Turks, and his family will be driven out of the country, as they could not oppose the Turks for any length of time should they really make up their minds to take over the country and employ force, and for this reason he is very eager to know whether His Majesty's Government would extend their protection over Katr in the event of a petition being submitted to them to that effect.[27]

In short, if Britain failed to protect Qatar as Shaikh Ahmed urged, the Ottomans would drive him out, along with the entire al-Thani family.

Despite the danger of the Ottomans absorbing Qatar in the absence of a foreign protection treaty, Gaskin informed Shaikh Ahmed that he was unable to reply as this subject was beyond his 'province'. However, Gaskin felt that the influence of al-Thani family on the tribes of Qatar would soon disappear entirely because most of the tribes were pearl fishers, who had grown rich through the generosity of Shaikh Jasim. Therefore, the al-Thani family needed British support to enable them to maintain their hold over the tribes and minimise the risk of losing Qatar.[28] Kemball, however, was not convinced with the Political Agent's reasoning. The acting Resident maintained:

> I am inclined to doubt if it is the case that the influence of the Thani family in Katr is rapidly waning. It may, however, be less than it was for the

161

reasons explained by the Assistant Political Agent. What these people fear is the extension of Turkish rule over Katr, as they know that they could not resist the Turks for long; and it is for this reason that the arrangement with His Majesty's Government is desired. If they were assured of the support of the British Government against Turkish encroachments, it is probable that the position of the family with regard to the tribes in Katr would be strengthened, and Sheikh Ahmed would have no necessity to appeal for support in asserting his authority over his own tribesmen.[29]

Therefore, the main reason for the al-Thani family's desire for British protection was to check the Ottoman expansion in the peninsula with British support.

CURZON'S HISTORIC VISIT

Preparations were being made for Curzon's historic visit to the Gulf. The Viceroy anticipated that Shaikh Ahmed might come to Bahrain to see him and state his views during his visit. Before he started his tour, Curzon telegraphed the India Office asking whether the British government had arrived at a final decision regarding British protection of Qatar.[30] The India Office and the Foreign Office discussed how to respond to Curzon's query. Finally, the Viceroy was informed that the British government was unable to conclude a fresh agreement with Shaikh Ahmed since, by withdrawing the mudirs appointed to Zubara and Wakra; the Ottoman government had complied with the British request for maintenance of the *status quo*. But the Shaikh should be assured of continued friendship so long as he abstained from entering into agreement with another power.[31] The British government's decision that no agreement should be concluded with Shaikh Ahmed was based on two main considerations. Firstly, such a treaty would constitute a disturbance of the *status quo*, which was held to be unjustified, given that the Ottomans had promised to cancel the appointments of mudirs in Qatar and secondly, Shaikh Ahmed would only use an agreement with the British government to shore up his diminishing influence.[32]

On 18 November 1903, a fleet of six British men of war and two Indian Marine vessels, one of them carrying the Viceroy, arrived in Gulf waters. This was Curzon's first visit to the Arabian Gulf and the first by such a high-ranking British dignitary. After a brief visit to Muscat the Viceroy's fleet arrived off the coast of Sharjah on 21 November 1903. On *Hardinge* he addressed the Trucial chiefs. Outlining British policy in the Gulf, the Viceroy announced:

> We opened these seas to the ships of all nations, and enabled their flags to fly in peace. We have not destroyed your independence but have preserved it. We are not now going to throw away this century of worthy and triumphant enterprise; we shall not wipe out the most unselfish page in history. The peace of these waters must still be maintained; your independence will continue to be upheld; and the influence of the British Government must remain supreme.[33]

Curzon not only recalled the establishment of British order and peace in the Gulf but also informed the Arab Shaikhs that the British were determined to preserve their paramount influence in the region through the preservation of Arabs' independence.

Following the conclusion of the meeting at Sharjah, the Viceroy's entourage arrived in Bahrain. Contrary to his expectations, neither Shaikh Ahmed nor any other member of the al-Thani family called on the Viceroy. Curzon felt no need to communicate with him despite his earlier enthusiasm. In fact, it was the British government which had ruled out the protection treaty. Having returned to India upon the conclusion of his Gulf mission, which also included a visit to the Shaikhdom of Kuwait, the Viceroy once again took up the question of the proposed Qatar treaty. In March 1904, reviewing the events in the Gulf, Curzon concluded that the problem in Qatar revolved around preservation of the *status quo* principle. But the Ottomans had not yet implemented that principle in Qatar despite their assurances. Therefore, Curzon sent a memorandum to the India Office, urging the government to reverse the decision:

> The status quo which His Majesty's Government desire to recognize and maintain includes the withdrawal by Turkey of any claim to administrative control or suzerainty over the peninsula; and it was contemplated that, provided

the Porte accepted the position and promptly withdrew the Mudirs at Wakra and Zobara, it might be undesirable to enter into any agreement amounting to a Protectorate Treaty with the Sheikh. The failure of the Turks to act up to their assurances in this matter, coupled with the intelligence received from Colonel Kemball as to the status of the Chief, seems to us to leave His Majesty's Government free to reconsider the matter; and we suggest that the time is now opportune for concluding an Agreement with Sheikh Ahmed substantially resembling those entered into with the Chiefs of the southern littoral, including the provision that he should not (1) enter into relations with, (2) receive the Representative of, or (3) cede territory to any foreign Power. We consider that it is desirable to consolidate such influence as we possess in El Katr, with special reference to the question of the pearl fisheries in the Gulf, as to which we have recently addressed you. The El Katr fisheries were extraordinarily productive in 1903.[34]

Although the Viceroy brought forward the question of the pearl fishery in Qatar, his main aim was to diminish the Ottoman claims on Qatar and sign a protection treaty with Qatar.

PEARLS AND THE QATAR TREATY
To protect the pearl fishery, Curzon was ready to retreat from the principal of *status quo*. The pearl banks off the coast of Qatar, the Trucial states and Bahrain, as well as al-Hasa and Kuwait, had since time immemorial been accessible without distinction to the Arabs of the entire littoral. Although there were no definite inter-tribal limits, the external boundaries of the fisheries were well known and therefore no chief had the right to grant concessions to outside parties. Since 1843, Britain had been forcing the Trucial chiefs to keep the peace at sea. In 1892, the Trucial Shaikhs, that is, the chiefs of Abu Dhabi, Dubai, Ajman, Sharjah, Ras al-Khaimah and Umm al-Qawain, concluded fresh compacts binding themselves not to enter into any agreement or correspondence with any power other than the British government and pledged 'not to consent, except with the permission of the British Government, to the residence within their territories of the Agent of any other state;

and not to cede, sell, mortgage, or otherwise give occupation of any portion of their territories, save to the British Government'.[35] In addition to these agreements with the Trucial chiefs, Britain also had been maintaining treaty relations with Bahrain,[36] which had for long been, to all intents and purposes, a British protectorate. British policy on Kuwait was however different from that of Bahrain. In the case of Kuwait, though the Ottomans had laid claim to certain ill-defined rights or sovereignty, Britain had always maintained that the Kuwaiti ruler enjoyed a large measure of practical independence and by means of a bond which amounted to an unofficial protectorate, the British government had gained control over the foreign policy of Kuwait.[37]

Qatar remained outside British jurisdiction because no formal agreement had been reached. The case for a treaty had become stronger than before in view of other European powers' interest for obtaining pearling concessions in the Gulf. For example, in 1903, two Frenchmen named Dumes and Castelin arrived in Bahrain and established an office there with the intention of prospecting the pearl banks round that Island.[38] As the 'El Katr fisheries were extraordinarily productive in 1903' and there was every possibility of foreign subjects taking advantage of this, Curzon and his colleagues in India considered it was desirable to consolidate British influence in Qatar and sign a protection treaty soon to safeguard the pearl fishery amongst other objections.[39] The India Office agreed with the government of India that in the absence of any formal convention with the British government, special attention should be given to the difficulties which might arise in connection with pearl fishing rights, which the tribes of Qatar possessed at the time. Therefore, St. John Brodrick, the Secretary of State for India in succession of Hamilton, strongly recommended to the Foreign Secretary, Lansdowne, that a way be found to conclude a protection treaty which would not only secure the peninsula from foreign interference but also assist the Indian authorities to protect the pearl fisheries in the Gulf.[40] Lansdowne sought O'Conor's views before taking a final decision. The Ambassador opposed the treaty. He felt that the delaying tactics by the Ottoman government in cancelling the appointment of the mudir of Wakra was sufficient justification for concluding a treaty with Shaikh Ahmed, as the Ottomans had already agreed to annul the appointment, albeit reluctantly. Nevertheless, such a move on the part of the British government might further disturb Anglo-Ottoman relations in the region; at the time Britain was insisting upon the removal of the Ottoman military post from

Bubian island (of Kuwait) as well as from Umm Qasr. However, if, after all, the British government decided to sign a treaty with Qatar on the lines of the Trucial convention, this should be done as quietly and secretly as possible, and be justified by Britain's need to gain the power to suppress 'piracy and maintain the pearl fishing rights of the Arab tribes on the coast'.[41]

The Foreign Secretary, Lansdowne, was at one with O'Conor's somewhat hesitating stance. Lansdowne, who was in favour of the preservation of the *status-quo* in the Gulf, informed the India Office that the Foreign Office was unable to give it blessing to the proposed treaty in view of Ambassador's opposition and the possibility of further exciting suspicion in the Ottoman government.[42] Brodrick was therefore unable to authorize Curzon to conclude an agreement with Qatar, despite the desirability of establishing a direct link with Qatar. Brodrick was well aware of Britain's disadvantageous position in the Gulf in the absence of such a treaty:

> The El Katr coast at present constitutes a break in the continuity of our maritime influence along the shores of this part of the Gulf, lying as it does between the island of Bahrein and the Pirate Coast. The absence of any agreement with the Sheikh may prove, in certain contingencies, a hindrance to the proper exercise by His Majesty's ships of their duties in the suppression of piracy and the maintenance of the peace of the Gulf. It also presents a difficulty of considerable importance from the point of view of the protection of the pearl fisheries from outside interference.[43]

The campaign for a Qatari protection treaty had foundered on the Foreign Office's failure to grasp the real benefit of such a treaty to the British chain of communication in the Gulf.

The Viceroy, however, was not one to give up easily. Major Percy Zachariah Cox, the Political Resident (April 1904 to December 1913), was instructed on 30 November 1904, to state his opinion on the proposed treaty with the Shaikh of Qatar.[44] Cox opined that the primary reason Shaikh Ahmed bin Thani wanted a protection treaty was that helping Britain to maintain order would likely strengthen his own position. Cox opposed Ambassador O'Conor's pleas for 'precautions and reservations', believing that such a policy would defeat its own aims and imply recognition of

Ottoman rights in Qatar for the first time. It would also prevent the British from obtaining a completely satisfactory agreement at a more opportune moment. The Political Resident therefore recommended that Shaikh Ahmed should be encouraged to give a written assurance to the Political Agent in Bahrain stating that he had succeeded to the rulership of Qatar following the abdication of Shaikh Jasim, due to the latter's old age, and that he accepted as binding the Agreement of 1868 concluded by his predecessor with the then British Resident, Pelly.[45] Curzon endorsed Cox's recommendation without hesitation, believing that the best way to deal with Qatar was by adherence to the Agreement of 1868, which would undoubtedly afford some measure of control over maritime relations between Shaikh Ahmed and foreigners.[46] Brodrick, endorsing Curzon's views, referred the matter to Lansdowne, stating that he would prefer to renew the Treaty of 1868, if a new treaty would be strongly opposed by the Porte.[47]

Lansdowne, surprisingly, declined to accept Brodrick's views, arguing that he saw no advantage in reviving the Agreement of 1868, as it was inadequate to serve British interests in Qatar. Furthermore, the Ottoman government had formally suppressed the mudirate of Wakra, depriving Shaikh Abdur Rahman of the title of mudir, as a result of the British government's repeated representations. The Foreign Secretary suggested that the matter should be deferred till the whole question of British policy in the Gulf had been fully examined by the Committee of Imperial Defence. Therefore, Curzon was told that the British government was unwilling to do anything that might disturb the *status quo* in Qatar, in view of the general sense of insecurity and suspicion in the neighbourhood of the Gulf.[48]

In spite of the shelving of the question of Qatar's protection treaty by the authorities in London, Captain Francise Beville Prideaux, who succeeded Gaskin as the Political Agent Bahrain in October 1904, produced a historical memorandum on Qatar, expecting that his views might be helpful to the Committee of Imperial Defence, which was examining British policy options in the Arabian Gulf region, particularly the proposed revival of the Agreement of 1868 with Shaikh Ahmed. The new Political Agent's research on Qatar and its affairs led him to conclude that the British government should modify its position on Qatar. He identified three policy options, one of which he favoured:

> First it seems possible to me that the suzerainty of
> the Chief of Bahrein might again be asserted over

the whole of the Peninsula excluding the Bida Chiefship, if our only reasons for prohibiting his interference in Katr affairs in 1875 A.D. was based on his weakness. The small seafaring tribes of the Katr coast are law-abiding people, and provided that they are not harassed, I do not think that they will raise any objections. Under no circumstances, would I permit any now uninhabited portions of the west coast, including Zubara, to be re-colonized, as its soil is unfertile and the pearl-banks being difficult of access from this tract, there can be no inducements except unlawful ones for people to settle there.

As the second alternative, the complete independence of the sea-faring tribes might be asserted, and in view of the fact that the Turks commenced the negotiations with the Sheikh of Bida in 1871 A. D., in direct disregard of their promises to observe the independence of the tribes with whom we had treaties, we might declare our intention of establishing a small post on the east or north coast of Katr for the prevention of piracies, and to maintain it so long as the Turks remain at Bida.

Thirdly, as we recognize that our interests in Katr Peninsula are purely confined to the maintenance of order along its coasts and on the adjacent seas, and that if the administration of the Turks was really effective we should not object to their establishing their dominion over those Arabs, who are not protected by anterior engagements with us, let us offer to recognize the sovereignty of the Porte over the whole of Katr on certain conditions. These, as it is impossible for the Sultan to guarantee an effective and humane administration at the present time, should be either that he should lease to us the whole of the coast from the southern frontier of Koweit, or even, including Koweit also, down to the northern frontier of Oman, extending in land all the way for five miles only, for a larger annual payment than the revenue which he at present receives, or that he should entrust the entire administration of this tract to British officers with full powers, to make all possible improvements, either arrangement to be in force for at least 50 years.[49]

Prideaux favoured the last of these three alternatives, which was based 'on a fallacious estimate and conception of the spirit' with which the Ottomans regarded British policy in the north of the Gulf and on the Arabian peninsula generally. The main reason for rejecting the first alternative was that Shaikh Isa of Bahrain was not in a position to exercise any jurisdiction in Qatar without substantial help and backing from Britain. However, Britain was unable to provide this unless a full British protectorate was declared over Bahrain, a move which had been considered outside the 'range of practical politics' as such a development would be an extra burden for Britain.[50] Prideaux's second option was also not practical unless the Porte was prepared to retire from the Qatar coast and leave Britain a free hand to control the lawlessness which then existed there, as Cox commented. Cox's approved in general Prideaux's historical memorandum on Qatar and submitted it to the government of India.[51]

PRIDEAUX'S MISSION AND INTERNAL ISSUES OF QATAR

The internal situation of Qatar became precarious due to Bedouin raids and counter raids. In an apparent move to restore law and order among the Bedouins on the south-western frontier of Qatar, in April 1905, Shaikh Ahmed led a raid by al-Murra and Bani Hajir (Mukhadhdhaba branch) against the Ajman and Bani Khaled (Amamarah) tribes in the districts of 'Jafura', close to al-Hasa. Achieving no major success, the raiding party, which lost five men, returned to Qatar.[52] Although, Shaikh Ahmed managed to impose order among the Bedouins on his south-western border, his relations with the Ottomans soon turned sour when he unwittingly murdered an Ottoman subject in September 1905, in al-Bida. On learning of his mistake, he expressed his regret to the Ottoman authorities and offered the usual blood money of Rs. 800, which was refused by the murdered man's relatives.[53]

While the Qatari-Ottoman dispute over the murder continued, Prideaux, the Political Agent, left Bahrain on 8th November 1905 for Qatar, accompanied by Enamul Haque, Agency interpreter, Mirza Abdul Latif, the Agency clerk, as well as a non-commissioned officer and an armed guard to collect information for the Persian Gulf Gazetteer, which was then under preparation, and to purchase a Qatari horse for British government service. On

169

the following day, Prideaux and his party arrived in al-Wusail, Shaikh Jasim's headquarters. There was nothing in the nature of a settled village in al-Wusail, which consisted of two or three mud huts behind a fort. However, about half a mile along the coast there was a large cluster of Bedouin tents 'whose occupants appeared to be more maritime than nomad in their pursuits judging from the number of pearling boats drawn up on the beach in front of them',[54] as the Political Agent observed. Prideaux also found two wells belonging to al-Wusail, which had been dug at the foot of a rocky hill about 1000 yards inland. A two storied watch tower stood on tope of the hillock.[55]

On 10 November 1905, Shaikh Jasim sent camels and donkeys to carry Prideaux and his team, from al-Wusail to Bu Hasa, 12 miles north-west, where Shaikh Jasim was staying at the time. The Political Agent was surprised to see the greenery and the agricultural development of Bu Hasa:

> Not a tree, however, did we see until at about 5 miles from the coast we surmounted a low ridge and came upon most refreshing and unexpected sight – a garden enclosed by a neat and low mud wall, 100 yards by 200 yards in area, and bordered by a line of tamarisk trees on all sides. Within were 3 masonry Persian wells of the largest size, worked by donkeys, and irrigating large plots of Lucerne grass as well as a number of pomegranate trees and some 300 hundred date palms.[56]

Bu Hasa oasis consisted of some 300 acres of low green land featuring coarse grass growing in raised clumps. It contained only one well, 30 fathoms deep, but the water was good.[57]

Upon his arrival at Bu Hasa, Prideaux met Shaikh Jasim and found him to be a 'patriarch of the ancient time' with a good knowledge and interest in politics despite his roughly 80 years. After conducting several interviews with Shaikh Jasim as well as collecting information from other persons for the invaluable Persian Gulf Gazetteer, to be written by John Gordon Lorimer, a British Civil servant, Prideaux returned to his boat at al-Wusail to sail round to al-Bida. The Agent and his staff landed in al-Bida after a journey of 22 hours. The Ottoman official who had been staying in the fort protested against the landing. However, Prideaux managed to make him understand the purpose of his mission in al-Bida, which involved no more than following up the commercial concerns of British subjects with Shaikh Ahmed.[58]

Having obtained Ottoman permission to move on, Prideaux met Shaikh Ahmed, who was 'a somewhat extraordinary character and at the same time an extremely astute man'. Shaikh Ahmed, 45 years old at the time, enjoyed much 'popularity and influence over his subjects'. And undoubtedly the people of Qatar and Bahrain regarded him as a strong man.[59] Having exchanged greetings, the Political Agent brought up the question of the protection treaty, asking whether there had been any change in Shaikh Ahmed's views about the desirability of a treaty with the British government similar to those of the Trucial chiefs. The adroit Shaikh replied that although he still wanted such a treaty, he could do nothing to bring it about so long as the Ottoman forces remained in al-Bida fort. Shaikh Ahmed wandered whether the British government would evict the Ottomans from al-Bida to achieve his long cherished desire. Prideaux replied that the British government was unable to evict the Ottomans from al-Bida, other than through diplomacy, in view of London's friendly relations with the Porte. The Political Agent was relieved that Shaikh Ahmed understood the real reason the Qatar protection treaty had failed to get off the ground. Having concluded his mission to Qatar, Prideaux returned to Bahrain on 19 November 1905.[60]

Shortly after Prideaux's visit to al-Bida, the internal political situation on the peninsula worsened following the murder of Shaikh Ahmed in December 1905, by his own Beni Hajir servant, the result of a personal grievance. The elderly Shaikh Jasim proved quite capable of dealing with the emergency, and once more took charge of the whole country. The elders of the Beni Hajir tribe, with whom the al-Thani maintained cordial relations, expressed their regret and assured Shaikh Jasim that they would hunt down and execute the guilty party. Accordingly, the Beni Hajir began to seek the murderer of Shaikh Ahmed and a few weeks later he was shot dead in Dhahran by one Bashir, a nephew of Salim bin Shafi, chief of the Makhadhabah division of the Beni Hajir, thus closing the feud arising from the death of Shaikh Ahmed. Although the foreign political relations of the peninsula were supposedly controlled by the military commander of the al-Bida fort,[61] Shaikh Jasim acted independently. For example, when Abdul Aziz bin Saud, son of the Wahabi Amir, paid a visit to the districts bordering Qatar in 1905, Shaikh Jasim sent the Amir a letter of welcome, with 8000 German Crown in cash and a present of rifles and rice and himself visited him at the wells of Araiq.[62]

CONCLUSION

At the beginning of the twentieth century the question of the British protection of Qatar dominated the Anglo–Qatari relations. While Shaikh Ahmed, wanted full protection of Qatar, the British authorities in the Gulf were interested only in protecting Zubara, not the whole of Qatar. This was due to the British policy of rejecting the Porte's claim to Qatar and reducing the chance of attacks being launched on Bahrain from Zubara. Curzon though, backed by Kemball, strongly advocated a protection treaty to serve British strategic interest in Qatar in order to establish a direct link with the peninsula and protect the rich pearl fishing belt off its coast.

Curzon failed to make significant progress with his plan to sign a treaty with Qatar because of the opposition of the British Foreign Office, supported by the Ambassador in Constantinople. As a result of these divergent views, the question of the protection of Qatar was shelved indefinitely. Anglo–Ottoman relations were further complicated by the Ottoman decision to despatch administrators to Zubara, Wakra and Khor al-Odaid. The administrators were later withdrawn against the background of growing British protests. The sudden demise of Shaikh Ahmed propelled the aged Shaikh Jasim back to the helm of administration in Qatar once again.

Notes.

1.   J. A. Saldanha, op. cit., Précis of Qatar Affairs; 1873-1904, vol. IV, p. 46.

2.   Ibid., p. 47.

3.   Ibid., p. 49.

4.   For full text see Aitchison, op. cit, p. 256.

5.   Gaskin to Kemball, no. 56, 22 March 1902, R/15/2/26.

6.   Gaskin to Kemball, unnumbered, 29 March 1902, ibid.

7.   J. A. Saldanha, op. cit., Précis of Qatar Affairs; 1873-1904, vol. IV, p. 45.

8.   Gaskin to Kemball, unnumbered, 29 March 1902; see also Gaskin to Kemball, no. 56, 22 March 1902, R/15/2/26.

9.   Ibid.

10. Kemball to the SGI/FD, no. 93, 26 April 1902, R/15/2/26.

11. SGI/FD to Kemball, no. 2012E, (Simla), 3 October 1902, L/P&S/10/4.

12. Kemball to the Officiating SGI/FD, no. 185, 27 December 1902, L/P&S/10/4.

13. J. A. Saldanha, op. cit., Précis of Qatar Affairs; 1873-1904, vol. IV, p. 50.

14. FS(Calcutta) to the Political Resident, unnumbered, 10 January 1903; see also the Political Resident to the FS(Calcutta), unnumbered, 27 February 1903, F.O.78/5380.

15. The V in C to the S/S for India, no. 83-E, 19 January 1903, as quoted in J. A. Saldanha, op. cit., Précis of Qatar Affairs; 1873-1904, vol. IV, p. 51.

16. Ibid, pp. 51-52.

17. Ibid, p. 52; see also Kemball to the SGI/FD, no. 47, 23 March 1903, F.O. 78/5380.

18. As quoted in J. A. Saldanha, op. cit., Précis of Qatar Affairs; 1873-1904, vol. IV, pp. 52-53.

19. Ibid, p. 53; see also Kemball to the FS(Simla), unnumbered, 28 April 1903, F.O. 78/6380.

20. Kemball to the SGI/FD, Enclosure no. 12 in no. 1, 28 April 1903, F.O. 78/5380.

21. J. A. Saldanha, op. cit., Précis of Qatar Affairs; 1873-1904, vol. IV, p. 54.

22. Ibid, pp. 54-55.

23. Kemball to the SGI/FD, Enclosure no. 18 in no. 1, 5 June 1903, F.O. 78/5380.

24. Kemball to the SGI/FD, Enclosure no. 23 in no. 1, 2 July 1903; see also F. E. Crew, Acting British Consul at Basra to Kemball, Enclosure no. 21 in no. 1, 9 June 1903; also see Gaskin to Kemball, Enclosure no. 17 in no. 1, (Extract), 1 June 1903, F.O. 78/5380.

25. Gaskin to Captain V. Dev. Hunt (First Assistant to the Resident), no. 161, 25 July 1903, F.O. 78/5380.

26. F. E. Crow to Captain V. Dev. Hunt, First Assistant to the Political Resident in the Gulf, no. 20, 12 August 1903, F.O. 78/5380.

27. Gaskin to V. Dev. Hunt, Enclosure no. 41 in no. 1, 20 Sept. 1903, F.O.78/5380.

28. Ibid.

29. Kemball to the SGI/FD, no. 33, 12 February 1904, F.O. 78/5380.

30. As quoted in J. A. Saldanha, op. cit., Précis of Qatar Affairs; 1873-1904, vol. IV, p. 55.

31. Lorimer, op. cit., Historical, vol. IB, p. 56.

32. Curzon, Thomas Raleigh and others to St. John Brodrick, no. 77, 31 March 1904, F.O. 78/5380.

33. As quoted in David Dilks, Curzon in India: Frustration, vol. 2, New York: Taplinger Publishing Co., 1969, p. 63.

34. Curzon, Raleigh and others to Brodrick, no. 77, 31 March 1904, F.O. 78/5380.

35. For full Text see Aitchison, op. cit., pp. 256-257.

36.   Ibid., pp.237-238.

37.   Colonel Malcolm John Meade to Shaikh Mubarak al-Sabah, unnumbered, 23 January 1899, R/15/1/472.

38.   SGI/FD to Brodrick, Enclosure no. 46, in no. 1, 10 March 1904, F.O. 78/5380.

39.   Curzon, Raleigh, and others to Brodrick, no. 77, 31 March 1904, F.O. 78/5380.

40.   The IO to the FO, no. 1, 18 May 1904, F.O. 78/5380.

41.   O'Conor to Lansdowne, no. 1, 28 June 1904, F.O. 78/5380.

42.   The FO to the IO, no. 2, 18 July 1904, F.O. 78/5380.

43.   Brodrick to Curzon, no. 41, 9 September 1904, F.O. 78/5380.

44.   The SGI/FD, no. 4042, 30 November 1904, R/15/2/26.

45.   Cox to the SGI/FD, unnumbered, 12 December 1904, R/15/2/26.

46.   Curzon to Brodrick, no. 4413, 30 December 1904, R/15/2/26.

47.   IO to the FO, unnumbered, 28 January 1905, R/15/2/26.

48.   FO to the US/S for India, unnumbered, 18 February 1905; see also the S/S for India to the V in C, unnumbered, 28 February 1905, R/15/2/26.

49.   Prideaux to Cox, no. 208, June 1905, R/15/2/26.

50.   Cox to S. M. Fraser (SGI/FD), no. 332, 16 July 1905, R/15/2/26.

51.   Ibid.

52.   See the Administration Report on the Persian Gulf Political Residency for 1905-1906, Calcutta (India): Office of the

Superintendent of government printing, 1907, (hereafter cited as Administration Report) p. 80.

53.   Ibid., p. 80.

54.   Political Agent Bahrain, to the Political Resident, no. 455, 23 December 1905, R/15/2/26.

55.   Ibid.

56.   Ibid.

57.   Ibid.

58.   Ibid.

59.   Ibid.

60.   For more details on Prideaux's Interviews with Shaikh Ahmed, see Prideaux to Cox, no. 457, 23 December 1905, R/15/2/26.

61.   Administration Report for 1905-1906, p. 81. See also Lorimer, op. cit., Historical, vol. IB, p. 826.

62.   Lorimer, op. cit., Historical, vol. IB, p. 835.

CHAPTER EIGHT

The Last Phase of Shaikh Jasim's Rulership and the Zakhnuniyah Crisis

The sudden demise of Shaikh Ahmed created political uncertainty in Qatar and immediately brought up the question of succession. Having taken charge of the country, Shaikh Jasim not only settled this vital political issue but also acted as mediator between Britain and bin Saud, who wanted a British protection treaty. While Qatar witnessed severe economic recession during the first decade of the twentieth century, both British and the Ottoman governments focused their attention on the internal matter of Wakra, which led to growing Anglo–Ottoman tension. However, the growing arms trade and the Ottoman occupation of Zakhnuniyah island once again brought up the question of the proposed Qatar treaty. Furthermore, the Zakhnuniyah crisis laid bare the power and functions of the kazi of Zubara, which extended as far as islands such as Hawar.

THE GOVERNORSHIP OF SHAIKH ABDULLAH
As mentioned in the previous chapter, Shaikh Jasim had no alternative but to assume leadership pending the selection of a suitable successor. The veteran Shaikh soon appointed Shaikh Ahmed's clerk, Ibrahim bin Saleh bin Bakr, as 'Amir al-Suq' (the bazaar master) for an interim period. In addition to the day-to-day administration of Doha, the bazaar master was responsible for settling criminal cases. However, the Ottoman garrison was responsible for punishing trouble makers and all important

political and tribal matters were referred to Shaikh Jasim, who was still living in al-Wusail.[1]

Having settled administrative and financial matters, Shaikh Jasim turned his attention towards the question of succession. Shaikh Khalifa, Shaikh Abdullah and Shaikh Abdur Rahman (sons of Shaikh Jasim), had been approached one by one but had all refused to take over. While Shaikh Khalifa and Shaikh Abdur Rahman both had their own duties in al-Bida and Wakra respectively, Shaikh Abdullah was more interested in pearl trading than government. The sons of the late Shaikh Ahmed were also considered by Shaikh Jasim, but they were too 'young and ignorant for the art of government'. Becoming chief of a society like Doha was no easy matter because of the presence of the Ottoman forces and tribal factions. The leader needed particular qualities, as defined by Shaikh Jasim himself:

> ...the ruler of Doha should at once be both a soldier and a statesman, able to beat out the tribes and to march long distances whenever necessary, and while in Doha to keep order in the town, to remain conciliatory with the different tribes and to keep himself out of playing in the hands of the Turks. The Turks according to Sheikh Jasim are powerless to do any harm when kept at a distance, but when close are difficult to manage.[2]

In the selection or election of the chief of Doha, the Ottoman garrison in al-Bida headed by Major Nemit Effendi, had no formal voice, voting power or power of veto. However, the delay in finding someone to take over political leadership became a cause of anxiety for the garrison commander, who now asked Shaikh Jasim to appoint a successor without further delay. Fearing the Bedouin threat to the security of the Ottoman overland mail link through Qatar to al-Hasa, the garrison commander also urged Shaikh Jasim to accept the Qaim-maqamship with a fixed salary, an appointment, which the Shaikh declined to accept.[3]

As pressure mounted, Shaikh Jasim intimated to the leading tribes of Qatar that they should elect a Shaikh as soon as possible. Although Shaikh Abdullah bin Jasim al-Thani, basically a religiously inclined man, initially declined the chieftaincy, stating that he was more interested in trade and commerce than administration, the tribes of Qatar eventually persuaded him to accept. Shaikh Abdullah became governor of Doha and heir apparent of Shaikh Jasim. His first task was to launch a series of campaigns against unruly tribes as far as Dohat as-Salwa and Oqair; by November 1906, tribal harmony had been restored between the people of Doha and the Ajman tribe.[4]

While Shaikh Abdullah restored tribal harmony, his relations with the British authorities in the Gulf soon became strained over the question of lawlessness on the Qatar coast. It was reported by the Resident that one Ahmed bin Salamah had arrived at Doha in July 1906, after attacking a boat on the pearl bank. His reported presence at Doha caused a good deal of anxiety among Bahrain's pearlers, though it was thought unlikely that he would commit any offence on the pearl banks east of Qatar peninsula.[5] As communication with the Ottoman authorities in al-Bida on this mater would imply the British recognition of the Ottoman authority there, the Political Resident, Cox, wrote to Shaikh Jasim at al-Wusail directly, asking him to take action against Ahmed bin Salamah and to instruct his dependents and Shaikh Abdullah to have no dealings with him. The Shaikh was reminded that it was Britain's responsibility to maintain peace and order on the waters of the Arabian Gulf:

> You are well-aware from of old time that the primary object of the British Government is to maintain absolute peace & security on the sea, so that all the natives of the littoral may trade without fear. If persons in the position of yourself & your sons harbour & entertain the evil doers how can there be peace! such conduct will bring trouble upon all.[6]

At the same time the British warship *Redbreast* was deployed on 10 September 1906, to reinforce maritime security around the Qatar pearl bank[7]. While the Ottomans protested against the presence of the warship, Shaikh Jasim denied the allegation that he had protected a lawbreaker, stating that neither he nor his son Shaikh Abdullah had entertained Ahmed bin Salamah, who had secretly entered the borders of the town. When Shaikh Abdullah's servants pursued him, he ran away in the night and stayed not a day. The British authorities in the Gulf decided not to pursue the matter further.[8]

SHAIKH JASIM'S SAUDI MEDIATION

In an apparent move to help to establish Anglo-Saudi relations on a solid foundation, Shaikh Jasim sent secretly an urgent letter to Prideaux through Sultan bin Nasir al-Suwaidi, head of the Sudan tribe of al-Bida, expressing his earnest desire to meet him at al-Wusail 'just for a quarter of an hour', without disclosing the agenda of the proposed meeting. Shaikh Jasim himself had written and signed this letter.[9] As the Political Agent was too busy with Agency work to visit al-Wusail, he sent Syed Mohammad Enamul Haque, Interpreter in the Bahrain Agency, to meet the Shaikh on 23 October 1906.[10] On the following day, Enamul Haque returned to Bahrain with Shaikh Jasim's verbal message, which dealt with the affairs of Abdul Aziz bin Saud.[11] It is important to note in this connection that in August 1906, bin Saud visited the Jafura desert, including the town of al-Urayq, a few miles east of Dohat as-Salwa, where Shaikh Jasim had a meeting with him, to the dissatisfaction of the Ottomans.[12] Although there is no record on the conversation between Shaikh Jasim and bin Saud at this meeting, it became clear that Shaikh Jasim had been asked to speak on behalf of bin Saud regarding British protection. Shaikh Jasim's verbal message, transmitted by Prideaux, reflected the agreement apparently reached at the meeting:

> The resources of Najd are stated to have been strained to the utmost by the recent internecine wars, and Bin Sa'ud considers that the oases of Hasa and Katif were always the most profitable possessions of his Wahhabi ancestors. He is anxious therefore to recover the two districts and he proposes that a secret understanding should be arranged between the British Government and himself under which he should be granted British protection from Turkish assaults at sea, in the event of his ever succeeding in driving the Turks, unaided, out of his ancestoral dominions. In return for this protection the Amir is willing to bind himself to certain agreements (probably similar to those of the Trucial Chiefs) and to accept a Political Officer to reside at his Court. The details of this secret treaty he wishes to be settled or discussed at an interview which he is ready to give me [Prideaux] either in person, or with his brother

representing him, at some convenient rendezvous in the desert.[13]

Shaikh Jasim's message was clear. Bin Saud wanted British naval protection to enable him to take possession of al-Hasa and Qatif and drive out the Ottomans. In return he was willing to sign a protection treaty with Britain. Shaikh Jasim wanted the reply to this vital message to be given to him 'only by word of mouth'; he would arrange to forward it to bin Saud by 'trusted messenger'. Prideaux, however, was against such a secret agreement with bin Saud in view of Britain's friendly relations with the Ottoman government.[14] The authorities in the Gulf therefore sent no immediate reply to Shaikh Jasim's plea. However, more than eight months later, Prideaux was instructed to tell Shaikh Jasim that the British government was unable to entertain his proposal for bin Saud's protection as British interests were 'strictly confined to the Coasts' of the Gulf.[15]

ECONOMIC CONDITION

The peninsula of Qatar faced severe economic hardship in 1907 due to the lack of rainfall. For want of grass for grazing and because of the extremely hot weather, most of the nomadic tribes of Qatar were obliged to send their sheep, camels and horses to the vicinity of al-Hasa and beyond.[16] Economic conditions in Qatar deteriorated further as the pearl trade declined, pearls being the only exportable item. Shaikh Jasim, who himself invested 1-2 million Indian rupees in pearls in 1907, sent his son Shaikh Abdullah to Bombay on 13 April 1907 to examine the pearl market personally. However, though Shaikh Abdullah stayed more than three months in Bombay, he managed to sell none of his own or his father's pearls, even after lowering the price considerably. The depreciation of pearls hit the bin Thani's income, so Shaikh Jasim established a customs house in Doha under the administration of one of his slaves, although the people of Qatar had been opposed to such a move previously, and had successfully resisted the attempt by the Ottoman government to establish a customs house at Doha in 1890. Their quiet acceptance this time around was perhaps due to their sympathy for Shaikh Jasim's 'pecuniary misfortunes'.[17] The customs house, however, failed to boost Qatari finances and the pearl market remained depressed throughout 1907 and 1908. The leading people of Qatar were all in a sense on the verge of bankruptcy, though this did not affect them as severely as those in other parts of the world, because their creditors among the general population had no effective means of redress.[18]

ANGLO–OTTOMAN ROW OVER WAKRA

While there was no sign of early economic recovery, Shaikh Jasim soon confronted with a new political issue, which once again drew both the British and the Ottoman to the environs of the peninsula. This was the discord between Shaikh Abdur Rahman bin Jasim al-Thani and al-Bu Ainain tribesmen of Wakra, which arose in September 1908, when the tribesmen refused to pay the annual tax levied on their pearling boats. Shaikh Abdur Rahman imposed of a fine of Rs. 10,000 on them and expelled six of their headmen from Wakra.[19] The Political Agent visited Shaikh Jasim at al-Wusail, who assured him that he would settle the matter amicably. Nevertheless, despite this assurance, the tribe appealed to the Ottoman Wali of Basra, Muharrem Pasha for protection, sending a fellow Bahraini-based tribesman, Ahmed bin Khatan as their envoy to request that an Ottoman military garrison be established in Wakra. The Wali wrote to Shaikh Jasim asking him to settle the dispute peacefully and asked the Wakra envoy to apply to the mutassarif of al-Hasa for assistance against Shaikh Jasim if the latter paid no heed.[20]

No sooner had F. E. Crow, British consul at Basra found out about the Wali's interview with the al-Bu Ainain's envoy, than he warned Muharrem Pasha that 'His Majesty's Government do not recognise Turkish sovereignty on the El Katr peninsula' and no Ottoman interference would be allowed in Wakra. The consul reminded the Wali that his intervention in the affairs of Wakra was contrary to the Porte's previous instructions to abolish the mudirate of Wakra and not to interfere in Wakra.[21] Crow's note, however, created uproar in Constantinople. On 25 July 1909, the Ottoman Ministry of Foreign Affairs rejected the British assertion:

> ...Ottoman sovereignty over the Qatar peninsula definitely exists. Jasim b. Thani, a Qatari citizen, in his official capacity as qaim-makam appointed by the Sublime Porte, is in charge of the administration of tribes and of ensuring safety and order in the place. He, together with the naib, his deputy, along with the administrators at Zubarah and Odeid and other nahiyes, are on the payroll of the Ottoman State. Jasim affixes his signature on

182

official papers in his official capacity of qaim-makam of Qatar. A flawless civil administration and the presence of a military power consisting of a battalion and two artillery units in Qatar are sound evidence of Ottoman sovereignty. Abdurrahman Effendi, son of Jasim, appointed administrator to the nahiye of Wakrah by decision of the administrative council, is now discharging his duty.[22]

The Ottomans were determined to resist British interference in Qatar, particularly in Wakra and considered Shaikh Jasim as the sole administrator of Qatar, despite his resignation of Qaim-maqamship.

Meanwhile, in December 1908, three members of the al-Bu Ainain tribe of Wakra, paid a visit to Mahir Pasha, Mutasarrif of al-Hasa, during the latter's short visit to Bahrain. The al-Bu Ainain delegation informed the Pasha that the members of the tribe in Wakra had settled their differences with Shaikh Jasim amicably. Expressing his satisfaction regarding their peaceful settlement with the bin Thani family, the Mutasarrif assured them that 'henceforth the tribe was under Ottoman protection' and they should refer to him in the event of their any encountering further difficulties with the bin Thani family. The al-Bu Ainain then abandoned their calls for the establishment of an Ottoman military post.[23]

Nevertheless, the British authorities in the Gulf took seriously Ahmed bin Khatan's mission to Basra to solicit the Ottoman aid on behalf of his kinsmen in Wakra against the al-Thani. Since Khatan was the subject of Shaikh Isa bin Ali al-Khalifa, chief of Bahrain, Prideaux at Cox's direction, warned the Bahraini chief stating that Khatan's act was highly inappropriate and he should take proper action against him and deter his subjects in future from either 'carrying the grievances of their kinsmen or encouraging the latter to represent them to the Turkish authorities or other foreign Governments'.[24]   While it was easy for the Political Agent to deter the Bahraini chief from entanglement in the affairs of Wakra, he could do little to prevent Ottoman involvement in the matter. Although, the Wakra crisis had been solved due to Shaikh Jasim's personal intervention, Prideaux worried that it might resurface again as a result of Ottoman interest and their backing of the tribe against Shaikh Abdur Rahman. Prideaux anticipated that the tribe might leave Wakra to avoid further direct confrontation with the governor Shaikh Abdur

Rahman. If this happened, the Ottomans, merely by threatening to intervene, would 'bring the bin Thanis to the point of conciliation where the tribesmen would meet them' thus making good use of the opportunity to increase their authority in the Qatar peninsula.[25] To avert such possibility, Prideaux wrote to Shaikh Jasim on 30 January 1909, expressing his desire to meet the Shaikh either at al-Wusail or Abu Dhaluf or Fuwairet.[26]

The choice of the north-west coast of Qatar rather than Doha or Wakra as the venue for the proposed meeting was intended to avoid attracting the attention of the Ottomans. Furthermore, the Ottomans were in an 'assertive mood' in the two latter places.[27] The two side thus agreed to hold the meeting in February 1908, at al-Wusail, Shaikh Jasim's headquarters. However, the meeting could not take place immediately: bad weather prevented Prideaux from travelling to al-Wusail.[28] On 8 March 1909, the Political Agent arrived at al-Wusail accompanied by one Yusuf bin Ahmed Kanoo. When Prideaux met Shaikh Jasim, the latter informed him that the al-Bu Ainain's problem at Wakra had been sorted out. The Political Agent drew Shaikh Jasim's attention to an alleged Ottoman plan to assume the administration of the Doha customs. To this the adroit Shaikh replied that the revenue from this source was very meagre, amounting to only Rs. 400/- per month. This was insufficient even to defray the cost of the bazaar watchmen. The Shaikh further stated that in order to satisfy the British government, he had introduced a heavy tax to kill the trade in slaves and arms. However, other necessary items were still allowed to enter Qatar untaxed. No sooner had the Political Agent returned to Bahrain, following the conclusion of the al-Wusail meeting, than an Ottoman official (Qadi) from the al-Bida garrison hurriedly left for al-Hasa via Bahrain to report Prideaux's al-Wusail visit to Mahir Pasha.[29]

The problem in Wakra resurfaced despite Shaikh Jasim's efforts to reach a permanent settlement. On 5 October 1909, following a quarrel with Shaikh Abdur Rahman, around 1000 male members of al-Bu Ainain tribe, headed by Abdullah bin Ali, left Wakra for Qasr-as-Subaih, a town on the mainland about 30 miles north of Qatif belonging to Shaikh Mubarak, the ruler of Kuwait; they did this with the blessing of both the Wali of Basra and Shaikh Mubarak.[30] Although initially Shaikh Mubarak extended all kinds of assistance, including supplying food to the tribe, his relations with them soon soured. This was due to the tribe's frequent visits to the Wali of Basra and the Mutasarrif of al-Hasa to request protection, and the Ottoman deployment of six soldiers at Qasr-as-

Subaih, the Ottoman authorities stated that the al-Bu Ainain were Ottoman subjects.[31] Shaikh Mubarak's relations with the tribe were further aggravated when the Ottoman government, in March 1912, appointed Abdullah bin Ali as headman of al-Bu Ainain and Mohammad bin Ali al-Bu Ainain as mudir of Qasr-as-Subaih with an allowance of TL5 per month.[32] However, the al-Bu Ainain eventually came to an understanding with Shaikh Mubarak, agreeing to pay him yearly allowance. In return for this payment, the Shaikh agreed to cease encouraging Bedouin to carry out raids on the tribe at Qasr-as-Subaih.[33] However, following the evacuation of the Ottoman forces from al-Bida in 1915, the tribe returned to their original homeland, reaching a compromise with the new ruler of Qatar, Shaikh Abdullah bin Jasim al-Thani.

ARMS TRADE

The economic and social conditions of Qatar improved in 1901, due to two successive bountiful pearling seasons.[34] The increased trade in arms and ammunition also contributed to the country's economic revival: it was stated in the British administration report that the Qatar coast, particularly Doha, had been a centre for the arms trade since 1909. Arms and ammunition were imported to Doha by land and sea from Muscat and then exported to the Persian coast and al-Hasa. Some local merchants of high standing were involved in the trade.[35] It should be remembered that ever since the tribal uprising against British rule on the north-west frontier of India in 1897, a portion of these tribes' armaments had been imported from the Arabian Gulf through the Persian territory of Jask. While the traffic in arms was prohibited in the British protected territories in the Gulf, the Qatar peninsula was wide open for such business in view of the absence of any legal treaty binding the peninsula to control the trade. The British authorities in the Gulf keenly felt the need for a treaty with Qatar when in December 1909, *Fox* captured a Persian Dhow, the *Khain*, carrying 419 rifles, 81 pistols and ammunition. The *Khain* had loaded the goods at Muscat with no flag or legal papers; part of the cargo was consigned to Qatar and part to merchants in Persia.[36] Thus Qatar became an *entrepot* for arms and ammunition.

OTTOMAN OCCUPATION OF ZAKHNUNIYAH

In March 1909, while attention was focused on the increased trade in arms, the Ottomans represented by the mudir of Oqair, occupied Zakhnuniyah, a small island 10 miles south of Oqair. The island had no permanent inhabitants but was frequented by the migratory Dawasir tribe during the winter season for fishing and hunting. The Zakhnuniyah fort, where the mudir hoisted the Ottoman flag, was almost in ruins; according to one account it was built by the Dawasir and later restored by the al-Khalifa. According to another account, the fort had been built by Shaikh Ali bin Khalifa (the father of the then ruler of Bahrain), around 1860, with the approval of bin Saud, the Amir of Nejd.[37] The mudir, claiming that the island belonged to the Ottomans, told the Dawasir whom he found there that they should recognize themselves as Ottoman subjects. The Dawasir, however, declined to do so, arguing that they might lose their possessions in Bahrain if they obliged. Having received this negative response, the mudir soon left the island.[38]

The occupation of Zakhnuniyah, which was also claimed by Bahrain, not only raised the question of British protection over Bahrain and the limits of Ottoman jurisdiction on the Qatar coast[39] but also constituted a clear breach of the *status quo* that Britain was determined to maintain. No sooner had the news of this occupation reached Bahrain than the Political Agent Prideaux, proceeded to the island, on 18 March 1909, to take stock. Prideaux found only one fishing boat on the beach; on the west side of the island he noticed an Ottoman flag rolled up and fastened tight to the mast. From Zakhnuniyah Prideaux proceeded to Hawar island, about one mile off the west coast of Qatar and which had been a 'dependency of the mainland state' (Qatar), since time immemorial. Here he found two winter villages consisting of 40 large huts belonging to the Dawasir and under the authority of one Isa bin Ahmed Dawasir. The Agent learned from the Dawasir chief:

> ...that Zakhnuniya was undoubtedly a possession of the Chief of Bahrain, but that the Dowasir regarded Hawar as their own independent territory, the ownership of this island having been awarded to the tribe by the Kazi of Zubara more than 100 years ago, in a written decision which they sill preserve.
>
> The contesting tribe named Al bu Tobais now in apparently extinct, but as the Kazi of Zubara was in those days an official of the Al Khalifa, the island would seem to be a dependency of the mainland State [Qatar], which the Chief of Bahrain still claims as morally and theoretically his.[40]

The Dawasir thus asserted that Hawar, unlike Zakhnuniyah, belonged to them and that they had derived this right from the Qadi of Zubara. Prideaux's reference to the Qadi of Zubara as an al-Khalifa official demonstrates his lack of knowledge of the system of administration and the role played by the Wahabis in reshaping the history of the region. It was the Wahabis (not al-Khalifa) who appointed a large number of Qadis (Judges) in the area stretching from Oman to Basra, including Qatar and Bahrain, during the period from 1787 to 1811.[41] In 1795 the Wahabis gained predominance in Zubara[42] and brought Bahrain under their direct control in 1801; this lasted until 1811 as discussed earlier. Isa bin Ahmed Dosari's statement to Prideaux in 1909, that the Dawasir's ownership of Hawar derived from the Qadi of Zubara, should be seen in this perspective. A century before, around 1808, the Qadi of Zubara was certainly not an al-Khalifa official but rather a Wahabi appointee.

Despite the historical and geographical links between Hawar and Zubara, Prideaux maintained that the Bahraini Shaikh should make a claim on Hawar just to avoid an Ottoman takeover of that island:

> I strongly deprecate letting the Turks keep Zukhnuniyah, as they will then naturally be encouraged to go on to Hawar, but if Shaikh Esa doesn't want or dare to assert his sovereignty over Hawar we shall be in rather a quandary.[43]

Nevertheless, despite Prideaux's urgings, Shaikh Isa made no claim on Hawar and on 30 March 1909, confined himself on Zakhnuniyah, only stating that the island belonged to him as his father had built the fort there fifty years ago and his subjects periodically residing there in order to fish.[44]

EVACUATION OF ZAKHNUNIYAH AND THE QUESTION OF THE QATAR TREATY

In 1909, four Ottoman soldiers were sent to Zakhnuniyah to guard the fort. As the British government had never recognised Ottoman jurisdiction as extending south of Oqair, the Viceroy of India recommended that the Shaikh of Bahrain's claim on Zakhnuniyah should be supported and the Porte be told categorically that the island belonged to Bahrain and that the garrison must be removed from the island.[45] Following the British representation to the Porte, the soldiers deployed to Zakhnuniyah were withdrawn to Oqair in early June, as Rifaat Pasha, the Ottoman Foreign Minister, confirmed to Ambassador Gerard Lowther, on 27 September 1909. However, to avoid disturbance of *status quo*, Lord Minto, the Viceroy of India was instructed not to allow the Bahraini Shaikh to hoist his flag on the island, as no flag had been flown since the time of his father.[46]. Captain Charles Fraser Mackenzie, the Political Agent at Bahrain (May 1909 to November 1910), persuaded the Shaikh of Bahrain not to fly his flag on the island, though some fifteen Dawasir men from Bahrain went to Zakhnuniyah in October 1909.[47]

The Ottoman evacuation of Zakhnuniyah was short lived. On 23 April 1910, it was reported that eight Ottoman soldiers had landed on the island and on every Friday the Ottoman flag was hoisted.[48] Lowther protested to the Porte as instructed by the British Foreign Office. While Lowther was awaiting to discover the result of his representations to the Porte, Mackenzie reported that a fresh party of Ottoman soldiers had been despatched from Oqair to Zakhnuniyah island.[49] Anglo-Ottoman relations deteriorated further when it was reported that in August 1910, a mudir was appointed to Khor al-Odaid, a town, which the Ottomans considered as belonging to them. The British maintained that it was within the territory of the chief of Abu Dhabi, who was under the protection of the British government. Khor al-Odaid was the western limit of the British sphere of influence, the nearest Ottoman garrison to Khor al-Odaid was at al-Bida on the Qatar peninsula, which was believed to consist of sixty Ottoman soldiers and four field guns at the time.[50]

Ottoman manoeuvres in Zakhnuniyah, which the Ottomans considered part of the mainland of Oqair, and their designs on Khor al-Odaid, the ownership of which was still in question, convinced Lowther that the Porte was determined to extend Ottoman sovereignty to the neighbourhood of Qatar. On 22 August 1910, in a long telegram to Sir Edward Grey, the British Foreign Secretary, the Ambassador pleaded that the British government should make up its mind which policy to adopt towards Young Turks' aggressive policy in the Gulf in general and Zakhnuniyah and Qatar in particular. Lowther added:

> ...the active forward policy of the Young Turk Vali [Wali] of Bussorah [Basra] and the Mutessarif of El Hasa have already brought us into sharp conflict, and there seems no doubt that we should insist on

Turkish exclusion from the district south of Ojair. If the Minister for Foreign Affairs, after studying the question of Zakhnuniyeh and Odied and consulting his colleagues, does not give categorical instructions for the non-interference of the Turkish local authorities, it would seem necessary, subject to the views of His Majesty's Government, to take a strong line. For, to the Turkish mind, Zakhnuniyeh is a sort of stepping-stone to El Katr, and perhaps even to the Trucial Coast. The Turks do not put forward any valid claims to justify their territorial acquisitions in those parts, but it is not difficult to glean that they base their claims on the fact that in the beginning of the sixteenth century a Turkish flotilla, under Piale Pasha, annexed Gwadur, in South Baluchistan, and sailed up the Gulf, compelling the Arab chiefs to acknowledge the sovereignty of the Ottoman Sultan and Caliph. They further feel that as the dominant Islamic power they have undefined right to bring under their allegiance and to protect the small Arab Moslem tribes, etc., in the Arabian peninsula.[51]

The Ambassador recommended a firm approach towards the new revolutionary regime at Constantinople, should they try to further strengthen their position in and around Qatar and attempt to assert control over the Trucial states, having established a foothold on the strategic island of Zakhnuniyah. Lowther urged that a comprehensive treaty be signed with Qatar before such things could take place:

In 1904, after the appointment by the Turks of a mudir at Wakra had been cancelled, the Government of India expressed forcibly their view that a treaty should be negotiated with Sheikh Ahmed, but this course was deprecated by my predecessor, Sir N. R. O'Conor, who regarded the conclusion of the proposed treaty as then inopportune. Should His Majesty's Government and the Government of India now decide on the expediency of making such a treaty, there would seem to me no objection from the Constantinople point of view, and the time may not be far distant when, as advocated by Lord Lansdowne in February 1905, a comprehensive, as opposed to piecemeal, treatment of outstanding questions in the Persian Gulf, including the withdrawal, absolute or against some quid pro quo, of the Turkish post at El Bida, may become imperative, if, indeed, it is not forced upon us.[52]

This change of attitude on the part of the Ottomans, as noted by the Ambassador, was due to the change of the government following the overthrow of Sultan Abdul Hamid II by a combination of army officers and young bureaucrats, in April 1909. The revolution was part of what is known as the Young Turk movement, the main aim of which was the salvation of the Ottoman empire and the restoration of the Constitution of 1876, curtailing the power of the Sultan.[53] The revolutionary regime brought a new dynamism to the politics of the Near and Middle East and began to emerge as a power to be reckoned with. Lowther was convinced that more Ottoman naval units would soon be cruising the Arabian Gulf, the fleet having been augmented by German built ships. As the presence of the more Ottoman ships in the Gulf would increase Ottoman influence, Lowther, therefore, advocated sorting out unfinished business with Qatar, the only country (apart from Nejd) which remained outside British treaty system of administration on the Arab shore of the Gulf.[54] However, the Foreign Secretary was still not inclined to take up the issue. The Ambassador was once again instructed to demand from the Ottoman government 'the immediate and permanent withdrawal' of the Ottoman garrison at Zakhnuniyah.

While the British Foreign Office referred Lowther's Qatar proposals to the India Office for study, the Ambassador, at the Foreign Office's instruction, lodged a protest with the Ottoman Foreign Minister Riffat Pasha, against the despatch of a fresh party of Ottoman soldiers to Zakhnuniyah from Oqair and the appointment of a mudir at Khor al-Odaid. The Ottoman Foreign Minister assured Lowther that he would soon reply regarding Zakhnuniyah and instruct the Wali of Basra not to make any appointment at Khor al-Odaid, or in the event of its having been made, to cancel it.[55] To frustrate Ottoman design in Khor al-Odaid, the British Foreign Office at the same time consulted with the Admiralty as to whether it would be possible to effect a landing at Khor al-Odaid to expel the Ottoman mudir, if the British government should decide to resort to naval force. The Admiralty recommended that the necessary action be taken by Indian troops, the role of the navy being limited to assisting in the landing operations, while the Royal Indian marine steamer *Hardinge* should facilitate any action which might be needed in connection with the blockade.[56]

While the preparation of a British landing at Khor al-Odaid were being made, on 28 September 1910, Rifa'at Pasha, the Ottoman Foreign Minister told Lowther categorically that the Sublime Porte had not appointed any mudir at Khor al-Odaid and had instructed the Wali of Basra, Suleyman Nazif Bey, to cancel any such appointment if already made by him and to abstain from any such appointment in future.[57] Later, on 17 October 1901, Rifaat Pasha assured Lowther that the Wali of Basra had been instructed that when the soldiers would left Zakhnuniyah at the end of the fishing season they should not be allowed to return to the island without prior orders from Constantinople. This Ottomans thus demonstrated an apparent willingness to return to the *status quo*. Furthermore, Lowther was told that 'no occupation from any other quarter would take place'.[58]

Contrary to the Ottoman Foreign Minister's assurances, local inquiries revealed the revival of the mudirates of Khor al-Odaid, Zubara and Wakra, although the new officials had not taken up their posts. Reacting sharply, on 21 October 1910, Grey instructed Lowther to make a further protest, making it clear that such appointments constituted disturbances of the *status quo* and reiterating that the Qatar peninsula was 'outside Turkish jurisdiction'.[59] The Ottoman Foreign Minister, however, stated that he was unaware of the alleged appointment of mudirs to these places and would investigate the matter.[60] To make a spot enquiry into the matter, the British ship of war *Redbreast* visited Wakra and Khor al-Odaid in November 1910, and found no sign of the Ottoman mudirs in those places. In fact, no mudirs had taken up their posts at Zubara, Khor al-Odaid and Wakra, where Shaikh Abdur Rahman remained in charge of the administration. The appointed mudirs were still residing at al-Hasa.[61]

The status of Zakhnuniyah island remained unclear. Although the Ottoman soldiers had left the island by the beginning of November 1910, the British authorities in the Gulf were convinced that their absence was temporary and might be an expedient devised by the Ottomans in order to avoid formally recalling their forces. The British expected a new force to arrive on the island soon.[62] Meanwhile, in order to try to ensure that the soldiers sent to the island would not return again following the end of the fishing session, F. E. Crow, British Consul at Basra, privately made contact with the Wali of Basra. To the great surprise of Crow, the Wali informed him that he had received no instruction yet on the matter of Zakhnuniyah, that the island was the Ottoman property and the British consulate had no right to enquire into Ottoman affairs either privately or officially.[63] The Wali's firm stand on Zakhnuniyah prompted the authorities in London to recommend deployment of the *Redbreast* around Zakhnuniyah island at once to remove any Ottoman soldiers or flags that might be found there. However, the Admiralty was unable to comply as the *Redbreast* had already left the Gulf of Arabia for Bombay.[64] The Zakhnuniyah crisis pushed London to take a final decision on Lowther's proposed Qatar treaty. The Resident in the Gulf had submitted his views on the treaty at the instruction of the government of India. Although, Resident Cox had previously advocated such a treaty with the Shaikh of Qatar as a bulwark against the Ottoman aggression, he now perceived certain difficulties. The Resident in fact wanted to link the Qatar treaty with an overall Anglo-Ottoman settlement in the Gulf. He put forward three reasons for this:

(1) Our agreement with Koweit should cease to be kept secret, and our intention to make the agreement effective should be intimated to the Turkish Government. I strongly recommend that anomaly of Turkish flag at Koweit should simultaneously be eliminated, if this can be done.

(2) Turks should be induced to confine themselves to their recognised possessions at Katif and Ojair [Oqair], Mudirates at Wakra, etc. should be finally abolished, and withdrawal of Turkish military posts from Jinnah, Um Kasr, Bubiyan, Zakhnuniyeh, and El Bidaa should be brought about us.

(3) Item No. (2) having been achieved, treaties should be concluded by us with Bin Thani, and, if necessary, with other headmen on the west coast of Katr, on the lines of the trucial coast agreements.

It is beyond my purview to deal with question whether coercive measures will be necessary to achieve the above ends, or whether it is possible to achieve them by means of comprehensive reciprocal compromise. I beg to say, however, that there is no quid pro quo in this sphere that I can conceive which could be offered to the Turkish Government in exchange for withdrawal of pretensions of Turkey. In some other sphere it might perhaps be possible to find such a quid pro quo.[65]

Cox was referring to the secret Anglo-Kuwaiti Treaty of 1899, under which Shaikh Mubarak al-Sabah agreed not to accept any foreign representatives or agents in his country nor to 'cede, sell, lease, mortgage or give for occupation' or for any other purpose any portion of his territory to any foreign power or the subjects of any power without the prior consent of the British government.[66] While Cox favoured disclosing the Kuwaiti treaty to the Ottomans, he wished for delay the conclusion of the Qatar treaty until the Ottomans withdrew completely from the region.

Therefore, by playing the Kuwaiti treaty card, Cox wanted to put pressure on the Ottomans to withdraw from al-Bida, Bubiyan island, Umm Qasr, Zakhnuniyah and Jinnah and confine themselves to their recognised territories at Oqair and Qatif. Ottoman withdrawal from the former places would undoubtedly help Britain consolidate its position in the Arab littoral states and protect the pearl fishery in the area. A general settlement with the Ottoman was therefore necessary before the conclusion of a treaty with Qatar. The Resident also drew attention to the urgency of the matter pointing out that if the Ottomans allied with Germany, Austria and Italy it would be more difficult to bring the Turks to an agreement in the Gulf. He added:

Achievement of above measures would completely consolidate our position on Arab coast in Gulf generally, especially in regard to pearl fishery question, and, if it is ever to be brought about, it seems essential that action should be taken now. If it is delayed, the suggested association of Turkey with triple Alliance and increase of her navy may render task very difficult, if not impossible. In any case, therefore, by one method or another, I respectfully urge necessity of adjusting position with Turkey forthwith and eliminating chronic elements of friction which have now become dangerous. On grounds above explained, I consider that it would be useless now to make treaty with Bin Thani, unless we are prepared to do so openly and to bring about elimination of Turkish influence from Katr, as suggested in second item above. Without such simultaneous action, existence of treaty would only be a source of danger to the Thani family from the present Turkish *regime*, and that they would themselves fight shy of it.[67]

The fate of the proposed Qatar treaty was therefore linked with an Ottoman withdrawal from al-Bida. Although a treaty with Qatar would have completed the British chain of defence and brought an end to the arms trafficking in Qatar, which was bristling with arms, the India Office endorsed Cox's recommendation without argument. In a memorandum to the Foreign Office, the India Office concluded that no treaty should be signed with the Shaikh of Qatar unless the Ottomans could be induced to confine themselves to their present recognised possessions at Qatif and Oqair.[68] The Foreign Office agreed entirely with the India Office and reiterated that it would be useless to conclude a treaty with Qatar.[69] Thus the proposed Qatar treaty was once again shelved.

CONCLUSION

Despite his old age, Shaikh Jasim was able to solve the internal political problems boldly and decisively, choosing a convincing heir apparent, the governor of Doha Shaikh Abdullah, who acted as Shaikh Jasim's deputy while financial crisis gripped the country. Shaikh Jasim's secret willingness to act as mediator on behalf of bin Saud, who wanted a British protection treaty similar to the British Trucial treaty, demonstrated the Shaikh's cordial relations with bin Saud and vice-versa. Even if bin Saud recovered al-Hasa and Qatif and drove out the Ottomans from these resource-rich places, Britain was not yet ready to accept Shaikh Jasim's plea for British protection of bin Saud from Ottoman assaults at sea. Britain's interests were limited to the Gulf littoral. Moreover, such protection would seriously disturb existing Anglo–Ottoman relations, as Britain recognized Ottoman jurisdiction at al-Hasa and Qatif. While Shaikh Jasim failed to achieve British protection for bin Saud, he skillfully and diligently solved the Wakra crisis minimizing the Anglo–Ottoman interference.

Ottoman occupation of Zakhnuniyah revealed the jurisdiction of the Qadi of Zubara, which extended even to the island of Hawar off the coast of Qatar. Most importantly, the Zakhnuniyah crisis brought up once again the question of the Qatar treaty. However, the divergent views of the Political Resident in the Gulf and the British Ambassador in Constantinople forced London to shelve the treaty until the Ottomans withdrew from al-Bida, the island of Bubiyan, Umm Qasr and Zakhnuniyah island. The final chapter reveals how Britain achieved its objectives in the Gulf, including the signing of a treaty with Qatar.

Note:

1. Prideaux to Cox, no. 468, 30 December 1905, R/15/2/26.

2. See 'Katar news' unnumbered, in Agent Bahrain to the Resident, 8 March 1906, see also Prideaux to Cox, no. 125, 9 March 1906, R/15/2/26.

3. Prideaux to Cox, no. 125, 9 March 1906, R/15/2/26.

4. Lorimer, op. cit., Historical, vol. IB, pp. 836-827.

5. Prideaux to Cox, no. 407, 10 September 1906, R/15/2/26.

6. Cox to Shaikh Jasim bin Thani, unnumbered, 25 September 1906, R/15/2/26.

7. Cox to Prideaux, no. 2264, 25 September 1906, R/15/2/26.

8. Translation of a letter from Shaikh Jasim bin Thani to Cox, unnumbered, 18 October 1906, see also Prideaux to Cox, no. 499, 16 November 1906, R/15/2/26.

9. See Translation of a letter from Shaikh Jasim bin Thani to Prideaux, unnumbered, 8 October 1906, R/15/2/26.

10. Prideaux to Shaikh Jasim no. 461, 21 October 1906, see also Translation of a letter from Shaikh Jasim to Prideaux, unnumbered, 24 October 1906, R/15/2/26.

11.   Prideaux to Cox, no. 500, 17 November 1906, R/15/2/26; see also Appendix no. 7.

12.   Administration Report for Bahrain for 1905-1906, pp. 81-82.

13.   Prideaux to Cox, no. 500, 17 November 1906, R/15/2/26; see also Appendix no. 7.

14.   Ibid.

15.   Cox to Prideaux no. 1320, 22 June 1907, R/15/2/26.

16.   Administration Report for Bahrain for the year 1907-1908, p. 95.

17.   Ibid, p. 101; see also translation of a letter from Shaikh Jasim bin Thani to Prideaux, unnumbered, 17 April 1907, R/15/2/26; for more on the decline of pearl trade see also translation of a letter from Shaikh Jasim to Prideaux, unnumbered, 2 August 1907, R/15/2/26.

18.   Administrative Report for Bahrain for the year 1908, p. 88.
19.   Ibid., p. 93.

20.   Ibid., see also Crow to Cox, no. 15, 1 December 1908, R/15/2/27.

21.   Crow to the Wali of Basra, no. 153, 1 December 1908, R/15/2/27.

22.   As quoted in Kursun, op. cit., p. 119.

23.   Political Agency Bahrain to the First Assistant Resident, Bushire, no. 669, 12 December 1908, R/15/2/27.

24.   Prideaux to Shaikh Isa bin Ali al-Khalifa, no. 59, 24 January 1909, R/15/2/27.

25.   Prideaux to Cox, no. 74, 30 January 1909, R/15/2/27.

26.   For full text see Prideaux to Shaikh Jasim, no. 73, 30 January 1909, R/15/2/26.

27.   Prideaux to Cox, no. 74, 30 January 1909, R/15/2/27.

28.   Prideaux to Shaikh Jasim, no. 107, 13 February 1909, see also translation of a letter from Shaikh Jasim to Prideaux, unnumbered, 3 February 1909, R/15/2/26.

29.   Prideaux to Cox, no. 207, 4 April 1909, R/15/2/25.

30.   For more on the migration of al-Bu Ainain tribe from Wakra to Qasr al-Sabyia, see First Assistant Resident Bushire Residency to Captain W. H. I. Shakespear, Political Agent, Kuwait, no. 2536, 5 October 1909, see also Political Agency Bahrain to Shakespear, unnumbered, 24 March 1909, R/15/5/86.

31.   See the extract from Bahrain News, no. 48, 20 February to 2 March 1912, R/15/5/86.

32.   Extract from Qasr al-Sabiya News, no. 58, 3 to 17 March 1912, R/15/5/86.

33.   Extract from Qatif News, Agent Bahrain to the Resident, no. 8, 3 March 1913, R/15/5/86.

34.   Administration Report for Bahrain for 1910, p. 86.

35.   Administration Report for Bahrain for 1911, p. 100.

36.   See the report on 'Arms Traffic in the Persian Gulf' Political Department, FO, no. B175, 10 June 1910, F. O. 881/9835.

37.   Prideaux to Cox, unnumbered, 20 March 1909; see also Prideaux to Cox, no. 207, 4 April 1909, R/15/2/25.

38.   Prideaux to Cox, unnumbered, 20 March 1909, R/15/2/25.

39.   IO to the FO, unnumbered, 26 May 1909, F.O.371/776.

40.   Prideaux to Cox, no. 207, 4 April 1909, R/15/2/25.

41.   Shaikh Abdullah Uthman Ibn Bashar, *Aunwan al Majd fi Tariq Nejd*, Riadh: Matba'at Daarath al-Maalk Abdul Aziz, 1982, p. 175.

42. *Lam' Al Shihab Fi Sirat, Muhammad bin Abd Al-Wahab*, British Museum, Manuscript Division, unpublished, 1816, pp. 202-203.

43. Prideaux to Cox, unnumbered, 20 March 1909, R/15/2/25.

44. See the translation of a letter from Shaikh Isa bin Ali al-Khalifa, chief of Bahrain to Prideaux, unnumbered, 30 March 1909, R/15/2/25.

45. The Viceroy of India to Viscount Morley, unnumbered, 23 May 1909, F. O.371/776.

46. Lowther to Grey, no. 1, 27 September 1909, also see the Secretary of State to the Viceroy of India, unnumbered, 13 October 1909, F.O.371/776.

47. A. P. Trevor, Assistant Political Resident to Captain Charles Fraser Mackenzie, Political Agent, Bahrain, no. 745, 20 November 1909, R/15/2/25.

48. See G of I to Morley, Enclosure in no. 52, 25 July 1910, see also Cox to the G of I, Enclosure no. 7 in no. 82, 16 July 1910, F. O. 881/9746.

49. Major Cox to the SGI/FD, no. 824, 29 August 1910, L/P&S/10/162.

50. Louis Mallet, FO to the Admiralty no. 141, 22 September 1910, F. O. 881/9746.

51. Lowther to Grey, no. 603, 22 August 1910, F. O. 881/9746.

52. Ibid.

53. For more details on the Young Turk Revolution see M. E. Yapp, The Making of the Modern Near East 1792-1923, London and New York: Longman, 1987, pp. 183-194.

54. Lowther to Grey, no. 603, 22 August 1910, F. O. 881/9746; see also Grey to Lowther, no. 312, 17 October 1910, L/P&S/10/162.

55. Grey to Lowther, no. 286, 26 September 1910, L/P&S/10/162; see also Lowther to Grey, no. 206, 28 September 1910, L/P&S/10/162.

56. See the Admiralty to the FO, no. 142, 23 September 1910; see also Malet to the Admiralty, no. 141, 22 September 1910, F.O.881/9746.

57. Lowther to Grey, no. 152, 29 September 1910, F.O.881/9746; see also S/S for India to the V in C, unnumbered, 4 October 1910, R/15/2/27.

58. Lowther to Grey, no. 206, 17 October 1910, L/P&S/10/162.

59. Grey to Lowther, no. 317, 21 October 1910, L/P&S/10/162.

60. Lowther to Grey, No. 72, 31 October 1910, F.O.881/9907.

61. Mackenzie, to Cox. No. 720, 29 October 1910, R/15/2/25; see also Charles M. Marling to Grey, no. 1, 23 November 1910, L/P&S/10/162.

62. Marling to Grey, no. 248, 14 November 1910, F.O.881/9907.

63. Crow to Lowther, no. 106, 27 October 1910, F.O.881/9907.

64. FO to the IO, no. 115, 6 December 1910, F.O.881/9907; see also the IO to the FO, no. 120, ibid; see also the Admiralty to the FO, no. 1, 21 December 1910, F.O.371/1004.

65. The IO to the Earl of Crew, Enclosure in no. 1, 1 December 1910, F.O.371/1004; see also the GI to the Earl of Crew, Enclosure in no. 120, 1 December 1910, F.O.881/9907.

66. For full text, see Aitchison, op. cit., p. 262.

67. The V in C to the IO, unnumbered, 1 December 1910, L/P&S/10/162.

68. See ibid; see also the Memorandum on 'The British relations with Turkey, in the Persian Gulf', by the IO, no. B181, 7 December 1910, L/P&S/10/162.

69.   Mallet to the US/S, IO, no. 14632/10, 14 December 1910, L/P&S/10/162.

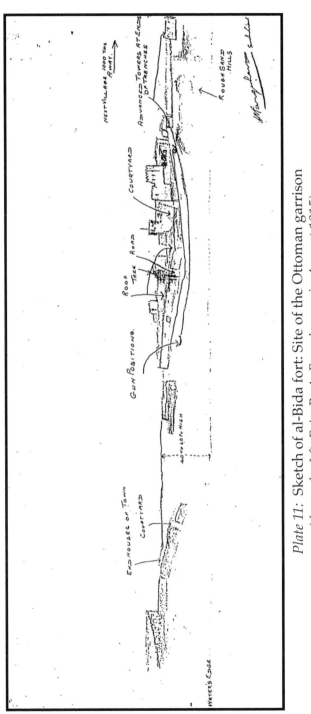

Plate 11: Sketch of al-Bida fort: Site of the Ottoman garrison
(drawn by Lft. Eric, R. A. Farquharson in August 1915)

Plate 12: Shaikh Abdullah bin Jasim al-Thani, ruler of Qatar (1913-1949), the first signatory to the Anglo-Qatari Treaty of 1916

Plate 13: Sir Percy Cox: The architect of the 1916 Treaty

CHAPTER NINE

The Baghdad Railway and the Status of Qatar

Having decided to settle all the Anglo-Ottoman disputes in the Arabian Gulf at the same time rather than on a piecemeal basis, Britain concentrated on the questions of Kuwait, Qatar and the Baghdad Railway project, in which Britain wanted equal partnership with other participating nations. Anglo–Ottoman correspondence and negotiations during the period from 1911 to the conclusion of the un-ratified convention in 1913, mainly focused on the status of Qatar and Kuwait and the ownership of the island of Zakhnuniyah. All the outstanding questions were resolved, including recognition of the independence of Qatar under Shaikh Jasim and his heirs, through the convention. The withdrawal of the Ottoman forces in 1915, bringing an end to the decades old occupation of al-Bida, was a landmark in the history of modern Qatar. The evacuation paved the way for signing of the Shaikh Abdullah-Cox treaty of 1916.

BRITAIN AND THE RAILWAY
Before discussing the Anglo-Ottoman debate on the status of Qatar, it is vital to examine the factors that led to Britain's involvement in the Baghdad Railway scheme, with which Qatar became linked. The Anatolian Railway Company was established in 1888 by two German financiers, Herr Kaula of the Wurtembergische Bank and Dr George Siemens, the director of the Deutsche Bank, to construct a railway line from Constantinople to Angora (present Ankara). On 25 November 1899, Baron Von Marschall, the German Ambassador

in Constantinople, persuaded the Ottoman government to issue a decree granting the Anatolian Railway Company a preliminary concession to construct a railway line from Konya to Baghdad.[1]

Germany's main objective in constructing railways in the Ottoman empire was to exploit the Ottoman markets as an outlet for its own rapidly growing industry. Sultan Abdul-Hamid II favoured the Baghdad railway project as he believed it would make a significant contribution to the economic development, unification and security of his Asian domains. The Sultan therefore permitted a commission of experts appointed by the Anatolian Railway Company to survey the route of a possible expansion of the projected Konya-Baghdad line to the Arabian Gulf. The commission, headed by Herr Stemrich, the German Consul-General in Constantinople, arrived in Kuwait in early January 1900, to identify a terminus for the planned Berlin-Baghdad railway. The commission selected Kadhamah at the head of Kuwait bay as a suitable site and attempted to open negotiations with Kuwaiti ruler, Mubarak bin Sabah, for its purchase. When Shaikh Mubarak refused to negotiate, the commission recommended its acquisition directly from the Ottoman Sultan.[2]

Britain was concerned that if the Germans established their railway terminus in Kadhamah, which was the best harbour in the entire Gulf and the most suitable for a military base, it would be difficult to dislodge them later, should this prove necessary. Therefore, upon the Foreign Office's recommendation, Lord Salisbury, now the British Prime Minister immediately agreed to disclose to Germany the nature of the existing Anglo-Kuwaiti Exclusive Agreement of 1899. Accordingly, on 10 April 1900, Sir Nicolas O'Conor, the British Ambassador to Constantinople told Baron Von Marschall, his German counterpart that Britain had an agreement with Kuwait which prohibited the ruler from making concessions to any other power without the prior consent of the British government. Von Marschall, however, refused to accept this, arguing that the agreement did not prevent construction of a Railway, as the Shaikhdom was still 'a part of the Ottoman Empire'. This unrealistic assertion compelled O'Conor to tell him that the Kuwaiti ruler was 'not a free agent'. Later that day, O'Conor saw Tawfiq Pasha, the Ottoman Foreign Minister, and told him that since Britain had 'certain agreements' with Kuwait, it would not allow any other power to assert rights or privileges over Shaikh Mubarak's territory. As the Pasha raised no objections the Ambassador believed that the Ottomans were already aware of the nature of the Exclusive Agreement of 1899. Despite these warnings, the Anatolian Railway Company made preparations for the construction of a pier at Kadhamah, putting the question of the international status of Kuwait in the spotlight. In June 1900, Sir Frank Lascelles, the British Ambassador in Berlin, clarified Kuwait's position to Chancellor Von Bulow informing him that Shaikh Mubarak had been enjoying 'a large measure of independence', having maintained no political or administrative links with the Ottoman state.[3] As a result of this diplomatic representation, the Anatolian Railway Company made no immediate attempt to continue with the project. However, in March 1903, a definitive concession was granted to the company, owned principally by the Deutsche Bank and French investors. The award of the contract paved the way for intensifying European rivalries in the Middle East and the Gulf.[4]

QATAR IN THE ANGLO-OTTOMAN NEGOTIATIONS

Definite change occurred in the British foreign policy following the conclusion of the Anglo–Russian convention of August 1907, which related to Persia, Afghanistan and Tibet. The convention removed British concern with Russian ambitions in the Gulf.[5] British policy in the Gulf was now concerned mainly with Germany and the Ottoman empire and with ensuring the preservation of the British interests in the Baghdad railway project and the Gulf in general. With this objective in mind, Britain decided to participate in the Baghdad railway project and buy a plot of land on the foreshore of Bandar Shuwaikh in Kuwait with a view to securing it for a terminus. On 15 October 1907, Britain signed the secret Bandar Shuwaikh Lease Agreement with Shaikh Mubarak. The Kuwaiti ruler agreed to lease in perpetuity the strip of land at Bandar Shuwaikh to the British government for Rs. 60,000 per annum, granting it the right to relinquish the lease at any time should it wish to do so.[6]

Having gained the ownership of Bandar Shuwaikh and its surrounding foreshore and all the lands and the foreshore in the direction of Kadhamah to a distance of two sea miles, the British Foreign Secretary Sir Edward Grey, felt ready to enter into serious negotiations with the Ottoman government to explain to them, the British interests and policy in the Gulf region.[7] With this in mind the British Foreign Secretary, on 18 January 1911, outlined a new approach and a new formula to deal with the Ottomans. Britain wanted to participate in the Baghdad railway project as an equal partner and settle the question of the status of Kuwait and Qatar within the broader context of the railway. Grey was willing to recognize Ottoman suzerainty in Kuwait provided the Porte agreed to joint Anglo-Ottoman control of the proposed railway terminus.[8] On 1 March 1911, the Ottoman government submitted a memorandum to Ambassador Lowther stating that the Porte was ready to use Kuwait as the railway terminus if Britain recognized Kuwait as an integral part of the Ottoman empire and agreed to internationalise the railway with a 40 per cent share for the Porte and a 20 per cent stake each for Britain, Germany and France.[9]

On 30 March 1911, the British Foreign Office produced a counter memorandum to deal with the Ottomans. The memorandum maintained that the southernmost limit of Ottoman jurisdiction on the Arabian littoral should be Oqair and that the Ottomans should abandon their claims on the coast south of Oqair, including those relating to Zakhnuniyah or adjacent islands. Furthermore, places like Zubara, al-Bida, Wakra and Khor al-Odaid of Qatar coast were regarded by the British government as being outside Ottoman jurisdiction.[10] Endorsing the memorandum, the Foreign Secretary indicated at a meeting of the 'Standing Sub-Committee of the Committee of Imperial Defence on the Persian Gulf', on 24 May 1911, that he was ready to be flexible with the Ottomans provided they agreed to a definite limit to the Ottoman sovereignty on the Gulf coast and renounced their claim on the Trucial coast. Grey added that while the Porte had put forward no claim to the Trucial coast, it had been 'pushing forward lately down towards El Katr, Zakhnuniyah and so forth'. The Committee was determined to limit Ottoman sovereignty in the Gulf particularly in those places in which Britain had an interest.[11]

Contrary to Grey's willingness to grant concession, the Committee in their final meeting of 14 July 1911, showed no sign of compromise. It maintained that the southernmost limit of Ottoman suzerainty should be Oqair in al-Qatif, and that the British government could grant no concession to the Ottomans in Qatar, owing to its proximity to the pearl fishing belt. The Committee concluded that the Ottomans should be induced to withdraw their military outpost from al-Bida and to renounce all claims to any place on the Gulf littoral south of Oqair including the British protected state of Bahrain.[12]

Following the conclusion of the Committee's meeting on 29 July 1911, the British Foreign Office sent a memorandum to Hakki Pasha, the Ottoman Ambassador in London, outlining not only the British position on the status of Kuwait and the Baghdad railway but on the whole of the Arab littoral of the Gulf, including Qatar. Regarding the railway, the memorandum proposed that the

Ottoman empire, Britain, France, Germany and Russia should each receive a 20 per cent share – the idea being to secure 60 per cent or a majority share for the three Entente Powers (Britain, France and Russia). The memorandum also stated that the Ottoman government should recognize Britain's absolute right to police the Arabian Gulf and should renounce its claims to Qatar and Bahrain. Britain was also prepared to recognize Ottoman suzerainty over Kuwait provided the *status quo* was guaranteed and the validity of the Exclusive Agreement of 1899 was recognized. Furthermore, the Ottomans must recognise the islands of Warba and Bubiyan as parts of Kuwait.[13] The Anglo–Ottoman preliminary conversation based on the Sub-Committee's recommendation, was held in the British Foreign Office. Negotiations were confined to the settlement of the Baghdad railway issue. However, the negotiations were soon halted because of the development of the Agadir crisis involving Britain and Germany, and the outbreak of the Tripolitan war between the Ottoman empire and Italy.[14]

Meanwhile, an agreement was signed between the Ottoman government and the Railway company to extend the railway only as far as Basra, thus ending (at least for the time being) the prospect of using Kuwait as a terminus. In March 1912, Britain opened formal negotiations with the Ottomans. The Ottomans were ready to agree to all British demands except those relating to Qatar, stating that the Ottoman withdrawal from al-Bida would be viewed within the Ottoman empire as a cession of territory; they suggested joint policing of the Gulf instead.[15] On 20 March 1912, the Ottoman negotiator, Djavid Bey, submitted a draft reply to the British Foreign Office stating that the Porte was prepared to grant a 25 percent stake in the railway project to all affected powers except Russia. The Porte requested that Britain should not insist on the sacrifice of Qatar or Zakhnuniyah.[16] Neither the Ottomans nor the British wished to compromise on Qatar. Sir Arthur McMahon, the Foreign Secretary to the Government of India was convinced that the British recognition of Ottoman sovereignty in Qatar would give the Ottomans a firm foothold in the Gulf. McMahon stated:

> Turkish ownership here [Qatar] will give them for the first time in history a foothold in the Persian Gulf with many potential possibilities of development, not only in the interior of Arabia but in regard to the Gulf itself, which will in time enable her to claim the right of intervention in Gulf matters of which we desire to retain a monopoly.[17]

Lord Hardinge, who succeeded Lord Minto as the Viceroy of India in 1913, was at one with McMahon and thought that it would be disastrous to allow the Ottomans to consolidate their position in the Gulf. Hardinge went a step further, stating that in view of the British refusal to recognise their claims of ownership, the Ottomans had no rights at all in Qatar, either dejure or defacto.[18]

Many months later on 15 April 1912, Tawfiq Pasha, gave the Ottoman official reply to the memorandum of July 1911, to the British Foreign Office, rejecting many of the British proposals. As anticipated, the Ottoman government wanted to institute dual policing of the Gulf and retain territorial sovereignty over Qatar. It also wanted effective sovereignty over the islands of Warba and Bubiyan and wished to replace the 1899 Anglo–Kuwait Agreement with a new convention. Further, the Porte was only prepared to guarantee Shaikh Mubarak's properties in Fao provided he accepted Ottoman citizenship. The Ottomans were ready to renounce their claim to Bahrain but not on Zakhnuniyah. The Porte rejected the suggestion that the Ottoman empire, Britain, France and Germany should each receive a 25 percent share in the railway and insisted on reserving the chairmanship and casting vote for itself.[19]

The Ottoman reply was thus harsher than expected. The Porte showed no sign of accepting Britain's point of view on Qatar and Kuwait. Underlining British dissatisfaction, Alwyn Parker, assistant head of the Foreign Office's Eastern Department, wrote 'a pretty stiff letter' to Djavid Bey, maintaining that no compromise would be made on Qatar and Kuwait, and stating that the Ottoman reply was incompatible with British thinking.[20] Sir A. Hirtzel, the Secretary of the Political and Secret Department of the India Office, adopted a still firmer stand on Qatar. Expressing his dismay, Hirtzel wrote in a private letter to McMahon that it would be a big mistake for the British government to allow the Ottomans to consolidate their position in Qatar. He added, 'bitter experience has shown that in the Gulf His Majesty's Government always take the line of least resistance otherwise the Turks would have been out of El Bida 40 years ago'. Djavid Bey however, on receipt of Parker's letter privately informed Hirtzel that 'the formal reply is only put up as something to bargain about.' To this, Hirtzel wrote back, 'do not, therefore, fall off your perch when you read it as we did!'.[21]

Hirtzel and Parker got together to examine the Ottoman proposal in detail. On 24 April 1912, they produced a joint minute listing their findings and views. Regarding the Baghdad railway, the joint minute maintained that the Ottoman proposal for the distribution of a quarter share to the Ottoman empire, Britain, France and Germany, but excluding Russia, would strengthen Ottoman-German participation in the railway project. As Russia would insist on participating in the project if the French did so, it was in the British interest to insist on Russian participation in the Baghdad–Basra section. However, Hirtzel and Parker were willing to accept an Ottoman as the president of the company to make the agreement more acceptable to the Ottoman parliament. The minute noted that while the Porte was ready to recognize British special relations with Bahrain, it showed no sign of compromise on Zakhnuniyah. Furthermore, the Ottoman stand on Kuwait would not serve British political and commercial interests in that Shaikhdom. They recommended that Britain should hold to its earlier proposals with regard to Kuwait. However, Britain could recognize Ottoman 'suzerainty over Koweit' and recognize the Shaikh as an Ottoman Qaim-maqam with authority over a specified territory, provided the Ottoman government agreed to maintain the *status quo* in other respects and accept the earlier British suggestions as mentioned above.[22]

Last, but not least, the Hirtzel-Parker joint minute noted that the most thorny question in the negotiations was the status of Qatar. The Porte was reluctant to deviate from its position, though one of the main reasons for initiating the negotiations was 'to eliminate the Turks from the Peninsula' once and forever. Enumerating the aims and objectives of the Ottomans in Qatar in particular and the Gulf in general, the joint minute concluded:

> It is evident from this proposal, and the proposal that they should consolidate themselves on El Katar, as well as from the general attitude of the Turkish delegates, that what the Turks are aiming at is, under the proposed settlement, not only to secure what they already possess on sufferance, but also to obtain a footing which they have never had before. Now whatever may be said from the point of view of law, history, or policy, for the former aim, there is clearly no justification for the latter, and His Majesty's Government, whose whole object is to get the Turks, as far as possible, out of the Gulf, can clearly, in no circumstances, acquiesce in it. We submit that this should be made quite plain, and it might be well to remind the Turks of Lord

> Lansdowne's statement in the House of Lords on the 5th May, 1903, that "we should regard the establishment of a naval base, or of a fortified port, in the Persian Gulf, by any other Power as a very grave menace to British interests, and we should certainly resist it with all the means at our disposal."[23]

The Hirtzel–Parker joint minute was thus a fairly uncompromising document. The Ottoman negotiator assured Parker privately that the Ottoman reply 'must not be regarded as their final word'. However Parker doubted that the Ottoman government was 'most likely to yield' on significant points.[24]

The Ottoman concept of a Gulf settlement thus differed significantly from that of Britain. The Foreign Office now considered the draft and concluded that the Ottoman presence in Qatar 'strictly defined' might not be a threat to the British interest and a *'quid pro quo'* might be achieved. Furthermore, if Britain made no compromise 'on Qatar it might lose on Kuwait', where its position was precarious despite the secret Anglo-Kuwaiti Treaty of 1899.[25] In fact, the Foreign Office believed that the Ottoman presence in Qatar was a price worth paying to retain Kuwait.

The India Office reacted angrily to the idea of sacrificing Qatar, maintaining that any Ottoman presence there was unacceptable and the Britain's share in the proposed Baghdad railway should not be less than 50 percent. It concluded that 'the Turkish garrison at El Bidaa – unrecognised for forty years, but nevertheless in possession – may, indeed, be regarded as a symbol of British policy in the Persian Gulf'.[26] For the Foreign Office Parker found the India Office's insistence that Britain should have a 50 percent stake in the railway, rather than the 20 percent he India Office itself had proposed in an earlier draft, confusing, disturbing and certain to sabotage the agreement. Parker believed an agreement with the Ottomans was essential for the following five reasons:

(1) Turkey's military striking power at Koweit is increased by the advent of the railway.
(2) Our own power is decreasing owing to our virtual naval withdrawal from the Mediterranean.
(3) Turkey may be driven into the German combination.

(4)   The private fortunes of the Sheikhs of Mohammerah and Koweit are derived from Turkish properties.

(5)   Our diplomatic and historical claim to Koweit is, to my mind, very vulnerable.[27]

Parker was willing to compromise with the Ottomans on Qatar provided acceptable terms could be obtained on Kuwait. Parker's eagerness was however qualified. Britain should strive to reach a settlement, but:

> ...only on conditions safeguarding our interests- all they want is some formal concession, we need give them nothing substantial- and if we recognised, say, their [Ottomans] suzerainty over El Katr and strictly defined its geographical limits, I think they would be satisfied, and that it would mean the difference of success and failure in the negotiations.[28]

The India Office struck to its gun and justified the demand for a 50 percent share, arguing that the Ottoman suggestion for 25 percent without Russia's participation was a rejection of the British offer of 20 percent. Lord Crewe, the Secretary of State for India, defended the India Office's stand, maintaining that no serious attempt had been made by Britain to obtain a 50 percent share. Crewe saw no reason to alter the British proposal on Qatar.[29] The government of India, endorsing the 50 percent proposal went a step further with regard to Qatar:

> We consider that our conditions as to El Katr contain no margin for surrender, representing as they do the irreducible minimum. As we have no confidence that any conditions imposed, however rigidly defined, will be adhered to by the Porte, we feel strongly the necessity of the complete removal from the peninsula of Turkish troops. No concession can be suggested by us that can be made with safety to interests of India and Great Britain, and rather than that any concession should be made to Turkey in the [Persian] Gulf we would prefer to see negotiations broken off. Strenuous opposition should be offered in our interests to any

step tending to consolidate Turkish position on Persian Gulf littoral.[(30)]

Therefore, the inflexible stance of the government of India, headed by Hardinge, was rooted in serious concern over British interests in the Gulf involving the security of India and the routes to India.

On 27 June 1912, an inter-departmental conference was once again held at the Foreign Office to tackle the divergent views of the India Office and the Foreign Office. Sir Arthur Nicolson, the Under Secretary of State for Foreign Affairs, chaired the conference and Sir R. Ritchie and Hirtzel represented the India Office. The conference was also attended by Sir H. Llewellyn Smith and George John Stanley representing the Board of Trade. The conference, dominated by Nicolson, concluded that there was no point in trying for a 50 percent share in the railway since both Russia and France would object. A smaller share would be acceptable provided the Ottomans renounced their territorial claims on Qatar and Bahrain with the exception of their al-Bida fort. In addition to this, the proposed railway line should stop at Basra, and no plans would be made for further extension. Although the conference did not touch on the status of Kuwait, it was later decided to allow the Ottomans to station an agent there on condition that they agree to the autonomous status for Kuwait with the removal of their forces from Warba and Bubiyan islands.[(31)]

A memorandum was eventually written expressing Britain's position on the outstanding issues and on 18 July 1912, Grey sent it to Tawfiq Pasha. The memorandum stated that Britain was now fully prepared to withdraw its request for participation in the railway from Baghdad to Basra provided a satisfactory agreement was reached on other issues such as Britain's absolute right to control Gulf waters, and the status of Kuwait and Qatar. Reiterating Britain's earlier representations on the Ottoman presence in Qatar, the memorandum reminded the Ottomans that Qatar had never been locally considered 'as forming part or as being subject to the influence of the Ottoman Empire'. In fact, the memorandum attached great importance to the Ottoman abandonment of the peninsula:

> His Majesty's Government can only reiterate their opinion that any lasting settlement between the two Powers must provide for the definite renunciation by the Ottoman Empire of the whole

210

of the peninsula of El Katr, including El Bidaa. They have consistently opposed any consolidation of Turkish authority, which first attempted to assert itself so recently as 1870, and they are bound to continue that opposition.[32]

Therefore, the fate of the proposed Anglo-Ottoman agreement depended entirely on favourable resolution of the Qatar issue. However, no immediate Ottoman response was forthcoming owing to the outbreak of the Balkan war in October 1912, in which the Ottoman empire was a principal party. Eventually, extension of the Baghdad railway beyond Baghdad towards Basra was abandoned.

THE CONVENTION OF 1913 AND THE AUTONOMY OF QATAR

In February 1913, the Ottomans felt free to talk about the British proposal. The Ottoman Grand Vizir Hakki Pasha arrived in London and opened serious negotiations with Parker and Louis Mallet, the assistant under secretary of state for Foreign Affairs and for several weeks the negotiators focused on the status of Kuwait and Qatar. Eventually, on 7 April 1913, the Ottomans agreed to withdraw their troops from al-Bida entirely, on condition that the peninsula remained autonomous. However, they declined to make any concession on Zakhnuniyah and requested British recognition of their rights over that island and the entire coast of al-Hasa, opposite to Zakhnuniyah.[33] The government of India cleared the way for Qatari autonomy, stating that the Ottoman proposal was acceptable on condition that the autonomy should mean 'complete renunciation of all claims to suzerainty on the part of Turkey' and Britain should have the right to conclude necessary agreements with Qatar to 'secure maritime peace or suppression of elicit arms traffic'. However, the government of India rejected Ottoman claims to Zakhnuniyah.[34] The India Office took a different view. Although British recognition of the Ottoman jurisdiction over Zakhnuniyah might create disputes in the near future, the India Office was inclined to compromise on Zakhnuniyah on condition that a satisfactory agreement was reached with the Ottomans on all other matters.[35]

Eventually, on 3 May 1913, a joint Foreign Office-India Office report was issued on the progress of the negotiations with Hakki Pasha concerning Qatar. The Ottomans were to abandon their claims on Qatar and withdraw their forces from al-Bida garrison. The British government would ensure that Bahrain did not try to annex Qatar. In return for these concessions the Ottomans were to retain Zakhnuniyah island, including the strip of coast between Zakhnuniyah and Oqair. The Ottomans were to compensate Shaikh Isa of Bahrain to the tune of £1,000 for abandoning his claim of the island of Zakhnuniyah. All these points, including a compromising formula on Kuwait, the railway and other Gulf issues were incorporated in a draft agreement signed by both the parties on 6 May 1913.[36]

While Shaikh Mubarak was informed of the territorial aspect of the draft convention concerning Kuwait,[37] Shaikh Jasim was kept in the dark despite the fact that Qatar was one of the major issues in the Anglo-Ottoman negotiations. Cox, who was well informed about the progress of the negotiations, felt the necessity of explaining the 'gist of the Anglo-Ottoman Convention' to Shaikh Jasim, 'so far as it concerns Katar and Bahrain and in order to see what we are likely to get him to accept in the direction of closer relations with us'. Cox was convinced that advance knowledge of the planned autonomy of Qatar with the assurance of British protection would encourage the Shaikh to accept the following British requests:

(i)   To receive an Agent of the British Government of India at Al Bidaa; (for the present a Mohammedan Agent of the type of the Residency Agent at Lingah, working immediately under the Political Agent at Bahrain).

(ii) To allow British Indians to reside in the Katar ports for trade, and to afford protection for their lives and properties.

(iii) Not to levy duty at more than 5 per cent ad valoreem.

(iv) To render co-operation generally for the preservation of the Maritime Truce, and in particular for the suppression of the Slave Trade, Piracy and the Arms Traffic.

(v)   To issue a prohibition against the import or sale of Arms and Ammunition in Katar territory except under special license.

(vi) To undertake not to allow foreigners including British subjects, to acquire monopolies or concessions or to acquire pearling rights or use diving dresses in his waters, without the cognizance and approval of the British Government.

(vii)   Not to admit the Agent of any Power other than Great Britain (and Turkey) and not to correspond with the officials of any other Power without the cognizance and concurrence of the British Authorities.

(viii) To receive and protect a British Post Office and Telegraph installation (as at Kuwait and Bahrain) whenever the British Government considers that the step is needed.[38]

Although Cox strongly recommended informing Shaikh Jasim about the settlement of the status of Qatar, the historic Anglo-Ottoman convention was signed on 29 July 1913, without the knowledge of Shaikh Jasim. However, Cox's eight points became the basis for the Anglo–Qatari Treaty of 1916.

The convention, inter alia, recognised Kuwait as an autonomous qaza of the Ottoman empire with defined boundaries and Kuwait's ownership of the islands of Warba and Bubiyan for which it had been struggling since the beginning of the twentieth century. The Ottomans received Zakhnuniyah island in return for a payment of £1,000 to Shaikh Isa of Bahrain.[39] Article 11 of the convention defined the status of Qatar:

II. – El – Katr.
Article II
The Ottoman sanjak of Nejd, the northern boundary of which is marked by a line of demarcation defined in article 7 of this Convention, ends on the south at the Gulf opposite the island of Zakhnuniyah, which belongs to the said sanjak. A line starting from the extreme end of the said gulf shall run due south to Ruba-al-Khali, and shall separate Nejd from the El-Katr peninsula. The boundaries of Nejd are marked by a blue line on the map annexed to this Convention (Annex V A). The Imperial Ottoman Government having renounced all their claims with regard to the El Katr peninsula, it is agreed between the two Governments that the said peninsula shall be governed, as heretofore, by Sheikh Jassim-bin-Sani and his successors. His Britannic Majesty's

213

Government declare that they will not permit the
Sheikh of Bahrein to interfere in the internal affairs
of El-Katr to infringe the autonomy of the country
or to annex it.[40]

The independence of Qatar was thus recognised, under the
leadership of Shaikh Jasim and his successors, and the Shaikh of
Bahrain was barred from interfering in the affairs of Qatar.
Although the convention of 1913 did not define the boundary of
Qatar, it separated Qatar peninsula from the Ottoman sanjak of
Nejd by a blue line; beginning on the Arabian Gulf coast to the
west of Qatar, opposite Zakhuniyah island and running due south
to the Rub al-Khali.

In this connection, it is important to note that even before
the signing of the convention, bin Saud took advantage of the
Ottoman preoccupation with international affairs and occupied the
Ottoman sanjak of al-Hasa on 12 June 1913, forcing the Ottoman
forces to evacuate to Basra.[41] In view of this changed situation, the
Ottomans at once increased their strength at al-Bida garrison to 250
men under the command of Abdul Ghafur Effendi.[42] Although
Shaikh Jasim was previously against the presence of any Ottoman
troops in his country, he welcomed the new Ottoman detachment
as soon as they landed in al-Bida. This was because of his fear that
bin Saud might move on Qatar in view of his increased activities in
the Eastern Arabia.[43] To eliminate Shaikh Jasim's 'apprehensions'
with regard to bin Saud's intentions and boost the former's morale,
Cox sought authority from the government of India to inform
Shaikh Jasim about the relevant portion of the Anglo-Ottoman
convention concerning Qatar, stating that the independence of
Qatar was duly recognised by the convention. Cox strongly felt
that Shaikh Jasim would not 'part with the detachment of Turkish
troops unless we give him an assurance that we will protect him
from molestation by Bin Saud'.[44]

POST-CONVENTION QATAR
Unfortunately, Cox could not convey his morale-boosting message
to Shaikh Jasim as the latter died of old age on 17 July 1913. His son
Shaikh Abdullah bin Jasim al-Thani succeeded him, in accordance
with the wishes of his father and of the key tribal leaders. The
Political Agent, Major Arthur Prescott Trevor (November 1912-
May 1914), was worried that bin Saud might take the opportunity
to make Shaikh Abdullah turn out the Ottoman garrison and force
him to accept various other impositions that Shaikh Jasim declined

to accept. Therefore, Trevor suggested that action should be taken on Cox's recommendation as mentioned above as soon as the period of mourning for Shaikh Jasim was over.[45]

As Trevor anticipated, bin Saud addressed the new Qatari leader, asking him to meet him and expel the Ottomans from al-Bida. Bin Saud's letter to Shaikh Abdullah alarmed Hardinge.[46] Endorsing the views of the authorities in the Gulf, Hardinge urged London to deal with bin Saud firmly. The Viceroy added:

> ...We deem it, however, our duty to place before His Majesty's Government the very strong objections to which, in our opinion, a rigid adherence to that policy in the changed situation is open. It is clear from his relations with Sheikh of El Katr, and the recent reports as to apprehension of the Trucial Chiefs that Bin Saud is becoming a more and more prominent factor in the politics of the Persian Gulf littoral and has thus brought himself within the sphere of our interest and influence. As long as Bin Saud confined himself to El Hasa he could be ignored with impunity. Such a policy, however, cannot in our opinion safely be pursued now that there is a possibility of his interference in El Katr, with the Trucial Chiefs, and possibly in Oman where rebellion may give him an opportunity for encroachment on Muscat territory. It seems to us therefore that we must now decide whether he is to be conciliated or estranged. To threaten Bin Saud, without further parley, with forcible expulsion if he attempts to interfere in El Katr, affairs cannot but have latter effect while if threats were ignored, we should be compelled to undertake military operations, a consummation which it is most desirable to avoid. Unless we come to some mutual and amicable understanding with the Amir who has hitherto been friendly dispositions the probability is that we shall force him to adopt an attitude of permanent hostility, which cannot but react unfavourably on the Chiefs of the littoral. We therefore strongly advocate not any definite treaty with the Amir, but a friendly interchange of views in which we should intimate to him that we expect him not to interfere with the

territories of Chiefs with whom we have treaty on other relations in the Gulf littoral.[47]

Hardinge was prepared to take charge of the guardianship of Qatar even before signing a protection treaty with Shaikh Abdullah, confining bin Saud in al-Hasa. However, he wished to manage bin Saud by tact and diplomacy.

Although conclusion of an Anglo-Qatari treaty became more urgent in view of the changed circumstances in the Eastern Arabia, Britain was still not ready for such a move. The government of India backed the delay:

> In consequence of the death of Shaikh Jasim and Abdullah's uncertain position it would seem premature to open negotiations for concluding treaty until such time as it is known whether new Sheikh can establish his position as successor.[48]

The death of Shaikh Jasim, the pioneer of the treaty, was thus the main reason for the delay. Hardinge himself preferred to defer all actions on the treaty negotiations until Shaikh Abdullah had 'shown himself capable of establishing himself as ruler of El Katr'. There was general agreement between the India Office and the Foreign Office on Hardinge's proposal.[49]

In line with this decision, Cox arrived in Doha in August 1913 to assess Shaikh Abdullah's position in Qatar and informed him about the withdrawal of the Ottoman troops from al-Bida under the newly signed Anglo-Ottoman convention. The new Qatari ruler informed Cox about his position with regard to bin Saud stating that upon the death of Shaikh Jasim, bin Saud wrote to him stating that 'he must get rid of Turkish garrison' if he wished to maintain friendship with him. To this Shaikh Abdullah replied to bin Saud that the 'garrison was quite innocuous' and that to expel them would only make the Turkish government hostile, but that he undertook neither to allow the garrison to be 'increased nor to let the Turks use Katr as a base'. Appreciating the British recognition of the independence of Qatar, Shaikh Abdullah assured Cox that he was ready to co-operate with Britain in the matter of arms traffic, provided Britain dealt with his own legitimate concerns. However, Cox's conversation with Shaikh Abdullah in Doha was inconclusive and convinced Lord Hardinge that the time still was not yet ripe for a treaty. The Viceroy recommended:

In the circumstances it is clearly desirable to avoid haste and to defer action as regards treaties for the present. Delay in the matter will also accord with wish of the Foreign Office that an interval should be allowed to elapse between publication of Anglo-Turkish agreements and conclusion of treaty with Sheikh. When the Sheikh's position is assured, our relations with Bin Saud defined, and El Katr affairs have had time to settle, we are of opinion that suitable opportunity might be taken to conclude treaty.[50]

Furthermore, the Viceroy felt that a treaty could only be signed if Shaikh Abdullah agreed to accept a British Political Agent, to suppress arms traffic and not to 'enter into relations with, receive representative of, or cede territory to any other power than British'. In return for these significant concessions Britain would assure him of its 'protection by sea and good offices' in the event of 'unprovoked attack by land'.[51] On 11 September 1913, the Political Resident with the backing of the government of India told bin Saud:

> … I have my Government's authority to assure you that provided you undertake on your part to abstain from all action calculated to disturb the status quo, or to create unrest, among the Arab principalities whose rulers are in relation with the British Government, including the principality of Qatar, the independence of which under the Government of the late Sheikh Jasim and his successors of the Bin Thani family has been recently recognised by the British and Turkish Governments, the British Government will continue to maintain the friendly relations which have been sustained in the past.[52]

Bin Saud therefore had to leave Qatar alone if he wanted friendly relations with the British government.

On 27 October 1913, bin Saud replied to Cox's communication, assuring him that he would 'abstain from all action likely to be opposed to views of British Government'. In view of this assurance bin Saud expected that the British government would consider the 'expediency of pressing' the Porte to remove the Ottoman garrison from al-Bida.[53] The British Foreign Secretary Grey, declined to do so until the Anglo-Ottoman convention had been ratified and became operative. Therefore, Cox was instructed to explain to bin Saud that Britain was unable to press the Ottoman to withdraw early from al-Bida.[54] However, despite Britain's repeated efforts the Anglo-Ottoman convention of 1913 remained un-ratified at the time of the outbreak of the First World War in August 1914.[55] Britain declared war against Germany, and the Ottoman empire joined the war on the German side in November 1914, completely disrupting the Anglo–Ottoman understanding in every sphere. Major Stuart George Knox, Deputy Political Resident (March 1914-November 1914), informed Shaikh Abdullah along with other Gulf rulers about British entry into the war.[56]

THE WAR TIME SITUATION

Britain's involvement in the European war naturally disrupted its engagement in Qatar affairs. Knox was worried that bin Saud might well utilise the wartime situation to his advantage by adding to his territory the coast of Qatar and other neighbouring countries. Knox underlined the potential impact of such expansionism and suggested how best to deal with it:

> Again if Qatar should fall into the hands of Bin Saud, the consequences for the Trucial Chiefs, commencing with Abu Dthabi, are bound to be most serious. They have been severely scourged this year by the ravages of plague and, on the top of that comes financial ruin, caused by this most disastrous War now raging in Europe.
>
> The only remedy for this state of affairs would appear to be, as has been already stated, an attempt to preserve Shaikh Abdulah bin Jasim bin Thani and show, in good time, by marks of our favour towards him, to Bin Saud that he must reckon on our strong opposition in the event of any intrigues tending to undermine the position of our prestige.[57]

Knox thus favoured supporting Shaikh Abdullah to the hilt in order to counteract any outside attack on Qatar and preserve British interests. He sought authority to despatch Captain Terence Humphrey Keys, Political Agent Bahrain (May 1914 to March

1916), to Qatar and to sound out Shaikh Abdullah's views on British policy towards bin Saud.[58]

As no immediate sanction was granted by London, Knox lost patience. His pleading on behalf of Qatar became more strident:

> I therefore make bold to solicit the permission of Government to proceed with the steps advocated in correspondence ending with my No. Cf.52 dated 21st July last, and also that the Political Agent, Bahrain, may be directed discreetly to pursue the policy of keeping in touch with events in Al Hasa and Bin Saud's subordinates in that province. It is even worthy of consideration, I venture to suggest, whether a warning might not be addressed to the Amir implying that Government has seen with much concern that there were serious discussions in the family of the late Shaikh Jasim bin Abdullah bin Thani; that the new Shaikh, Abdullah, the eldest son and lawful heir of the late Shaikh, had, in accordance with universal custom, succeeded to the Shaikhdom; that his maintenance in that position and the independence of the Shaikh of Qatar were objects of their solicitude and that they hoped that any malcontents of his family who were short sighted enough not to perceive that their true interests lay in the enhancement of the dignity and position of the head of their family, would receive no encouragement at the hands of the Amir or from his officials.[59]

In short, Knox wanted a treaty with Shaikh Abdullah to strengthen Shaikh Abdullah's position at home and abroad and improve British relations with bin Saud. The Deputy Resident's request received a favourable response from London. The India Office was, as before, full-square behind a treaty with Qatar, leading the British Foreign Secretary Grey to clear the way for its implementation. His decision was communicated to Crewe just before the outbreak of war with the Ottomans:

> ...on the subject of the conclusion of a Treaty with the Sheikh of Katr, I am directed by Secretary Sir Edward Grey to state, for the information of the

Marquess of Crewe, that he concurs in His Lordship's proposal to authorise the Government of India to conclude at their discretion a treaty of the nature indicated and on the conditions mentioned in the event of war with Turkey occurring before the situation has been discussed with Bin Saud.

Sir Edward Grey also agrees that discretion might at the same time be given to the Government of India to instruct Captain Shakespear to warn Bin Saud.[60]

With the Ottoman entry into the war, nothing now stood in the way of a treaty with Qatar. The matter was left at the Viceroy's 'discretion whenever circumstances appear favourable to negotiate treaty with Katr Sheikh which would lead up to prohibition of arms traffic'.[61] Stamping out the arms trade in Qatar was one of Britain's key objectives in the treaty.

The government of India, headed by Viceroy, was, however, now in no hurry to conclude the treaty. The war in Europe led to a British naval blockade of the Arabian Gulf, which caused a decline in the arms traffic; this rendered the conclusion of a treaty with Qatar less urgent. Admitting the reasons for a treaty with Qatar, the government of India was not ready to proceed unless the Ottoman forces were driven out from the peninsula and Britain had concluded a treaty with bin Saud:

> Admitting, however, that His Majesty's Government have good reasons for pressing the conclusion of the Treaty with El Katr, the Government of India are of opinion that there are considerable difficulties, which make such a course undesirable at present. In the first place, a necessary preliminary to any negotiations with El Katr is the ejection of the Turkish garrison at El Bida. It is understood that a ship of war cannot be spared for the purpose, and it would be inadvisable to detach even a small force from Mesopotamia. Apart from this, the question of El Katr is intimately connected with the negotiations at present in progress with Bin Saud and it would be a grave error to prejudice those negotiations by concluding beforehand a Treaty with El Katr, the

terms of which Bin Saud might conceivably consider objectionable.[62]

Therefore, the government of India imposed two pre-conditions on the Qatar treaty: the removal of the Ottoman garrison and achievement of a settlement with bin Saud. The views of Cox, who resumed his duty in November 1914, was sought as to whether the conclusion of a treaty with Qatar was 'immediately feasible, and if so, with whom and on what terms it should be concluded'.[63]

OTTOMAN EVACUATION OF AL-BIDA

As the war progressed, events in Qatar took a dramatic turn. On 11 August 1915, news reached Bushire that a large number of Ottoman troops had deserted, diminishing the al-Bida garrison. Some of the deserters found their way to Tangistan on the Persian coast.[64] Keyes, the Political Agent Bahrain, was instructed by Cox to leave Bahrain on 18 August 1915, on board *Pyramus*, accompanied by *Dalhousie*. He arrived off al-Bida on 19 August 1915 and oversaw the flight of the remainder of the Ottoman troops. Having ensured that the historic evacuation was completed, on 20 August, *Pyramus* sent the following telegram from Doha:

> On threatening to land troops from Dalhousie the fort and trenches were evacuated by the Turkish detachment during the night. This morning an armed party was landed without opposition and seized store (s) which included one mountain gun two field guns. The breech blocks of these guns had been removed. Besides tents there were 14 rifles, 120 cases ammunition and 500 projectiles. For the field guns no powder was found. The ammunition and rifles and part of the military stores were on the advice of the Political Officer given to the Sheikh of Qatar who has given us every assistance. The projectiles, tents and military stores were destroyed as all Tangistani dows have for some time been in hiding but the guns were taken on board H. M. S. Pyramus.[65]

The demoralised Ottoman forces fled al-Bida fort, bringing to an end more than forty-three years of Ottoman occupation of the

221

town. The issue of the withdrawal of the Ottoman garrison was thus resolved.

THE TREATY NEGOTIAITONS

Because circumstances had changed and because Shaikh Abdullah had demonstrated good will and a readiness to cooperate with Britain, Cox felt in August 1915, the time was right to conclude the long discussed Anglo-Qatari treaty.[66] He suggested to the government of India that the draft agreement, already drawn up by Knox in July 1914[67] should omit all reference to the arms traffic, as this had dwindled almost to nothing. No specific reference should be made to posts and telegraphs as this might create difficulties for Shaikh Abdullah. Cox wanted to depute Colonel William Grey, the Political Agent Kuwait (January 1914 to June 1916) to proceed to Qatar at once to negotiate the treaty.[68] Hardinge, Viceroy of India gave his approval to Grey's proposed mission to Doha and agreed to drop the arms traffic clause. He was however unwilling to abandon the post and telegraph clause without a fight, as he believed it essential to establish post office and telegraph infrastructure in Qatar.[69] Crewe, the secretary of state for India, objected to dropping the arms traffic clause:

> Meanwhile, in view of uncertain future and strong probability that there will always be smuggling on a larger or smaller scale, I suggest for your consideration that it is more advantageous to include prohibition of arms traffic now than to leave Katr the one open spot in Gulf with possibility of having to make separate treaty later in less favourable circumstances.[70]

Arms traffic thus remained an issue in the Qatar treaty negotiations despite its marked decline in importance.

The British chief negotiator left Bahrain on 19 September 1915, on HMS *Lawrence* accompanied by Keys, the Political Agent Bahrain, to negotiate the treaty. Shaikh Abdullah was informed in advance of their mission. Cox was supposed to conduct the negotiations but was preoccupied with wartime military strategy in Iraq. Grey was thus deputed with full authority to sign the treaty on behalf of Cox, subject only to later ratification by the British government.[71]

Upon his arrival in Doha on 20 September 1915, Grey opened negotiations with Shaikh Abdullah on the British draft, which consisted of ten articles. Negotiations were initially confined to the articles I and II of the draft treaty, dealing with the maintenance of the maritime peace by Shaikh Abdullah and the British obligation to grant the same immunities and privileges to the Shaikh and his subjects which were enjoyed by the British friendly Shaikhs of the Trucial states and their subjects. Shaikh Abdullah raised no objections to the inclusion of these provisions. The negotiations then turned to the draft article III dealing with the arms traffic. Grey pointed out that the British government was working to suppress the arms traffic in the Gulf and made it clear that Shaikh Abdullah could no longer obtain arms from Muscat; Britain would, however, make the necessary arrangements to supply the Shaikh personally and the state of Qatar with arms. The Shaikh agreed that the trade was practically dead and informed the British negotiator that when some Afghans had visited the Doha port recently to buy rifles he had sent them away empty handed. Grey considered the matter settled. As soon as the arms clause was agreed the negotiators proceeded to article IV of the draft, concerning the admission of British subjects to Qatar for the purpose of trade and the appointment of a British agent at al-Bida for the transaction of such business. Shaikh Abdullah declined to accept the proposal, arguing that the Qatari people would oppose allowing in foreign traders because of their previous experiences. He was also opposed to the presence of a British agent in Qatar to 'guide' him. As Shaikh Abdullah remained firm on the issue, Grey moved on to article V. Shaikh Abdullah refused to accept the trade clause, arguing that the customs import duty upon British goods should not exceed that levied on his own subjects and should not exceed 5 percent *ad valorem*. However, Shaikh Abdullah agreed to articles VI and VII, which stated that he would not receive the agent of any other power and grant the pearl fishery concessions or any other concessions without the consent of the British government. Grey then moved on to article VIII, dealing with postal and telegraph infrastructure. Appreciating the benefits of such installations, the Shaikh explained the difficulty of accepting these articles in view of the involvement of foreigners in running these offices. Negotiations were suspended when the Shaikh became weary.[72]

The next day the negotiators considered articles IX and X of the draft treaty dealing with the limits of British protection of Qatar and British granting of good offices in the event of emergency. Lengthy discussions took place on these two vital clauses. While Grey offered Shaikh Abdullah, British protection of Qatar against aggression by sea, the Shaikh demanded protection by land too. Grey pointed out that it was beyond his power to make such a commitment. However, he pointed out that the extent of British support would depend upon that of Shaikh Abdullah's confidence in Britain and 'the loyalty or otherwise with which he carried out his obligations'. Shaikh Abdullah accepted this explanation with satisfaction. Grey then brought up the other unresolved clauses, particularly relating to the admission of British traders. Shaikh Abdullah once again declined to accept this or the other two disputed clauses.[73]

Grey thus failed to bring Shaikh Abdullah round to his views. Three clauses still remained unresolved. These were article IV (acceptance of British traders and an agent), article V (limitations of duty on British goods) and article VIII (establishment of a post office and a telegraph office).[74] Having failed to conclude the treaty, on 26 October 1915, Grey recommended that the government of India should accept Shaikh Abdullah's explanation of his inability to entertain British merchants in Qatar for the time being, and to negotiate the remainder of the draft treaty on the understanding that a subsidiary agreement regarding the clauses omitted would be made later.[75]

Grey's views were widely accepted by Cox, who now strongly favoured signing the treaty on the Shaikh Abdullah's terms. The Resident explained:

> I would recommend another attempt to get the Treaty signed without alteration in order to keep it in line with those in force in the case of the other Sheikhs in treaty with us, but I think we might give Sheikh Abdullah a separate letter at the same time explaining that it was important to have the treaty worded on the same lines as others but assuring him that we have no intention at present of pressing for fulfillment of Articles IV, V and VIII. As long as our right in each of these respects is clearly enunciated, I think we can afford to reserve them for the present for the reason that if trade and harbour facilities progress at Bahrain, as they are doing, the trade for the greater part of the Qatr peninsula will depend upon Bahrain, when we shall have practical control over it.[76]

Cox's strategy was to keep the disputed clauses in abeyance for an indefinite period in light of Shaikh Abdullah's objections and get the treaty signed. The Indian government agreed with this approach. In a telegram to the India Office, the Viceroy urged that Shaikh Abdullah's objections be accepted to save the treaty from collapse:

> Articles to which Sheikh takes exception are of no great importance at present time and by accepting remaining articles Sheikh undertakes to do practically all that we immediately require. We therefore propose to authorise Cox to negotiate Treaty on lines acceptable to Sheikh should he find any difficulty in securing Sheikh's acceptance of Articles to which he now objects. We consider it more important to secure a reasonably satisfactory Treaty than run the risk of a breakdown of negotiations by insisting on conclusion of a Treaty on all fours with other Trucial Treaties. Kindly telegraph your approval.[77]

Britain sought a treaty with Qatar partly because it was the only state on the Arabian shore of the Gulf remaining outside the British treaty system. The seven Trucial states and Bahrain had already signed treaties. Also treaties with bin Saud and with Kuwait were signed, although these were not the same style.

THE TREATY OF 1916

London also favoured a treaty with Qatar. Backing the Viceroy's recommendation, the India Office, with the agreement of the Foreign Office, instructed Cox to proceed to Doha and resume negotiations with the Shaikh.[78] HMS *Lawrence* arrived in Doha harbour on 2 November 1916 carrying Cox, Captain Trenchard Craven Fowle, Acting Political Agent in Bahrain (July 1916 to February 1918), and Haji Yusuf Kanoo, a prominent merchant and Shaikh Abdullah's commercial Agent in Bahrain. On the following day, Cox met Shaikh Abdullah and opened the negotiations on the three disputed clauses. The Resident's forceful and persuasive personality dominated the course of the laborious negotiations. Shaikh Abdullah eventually agreed to British desiderata on the three clauses in question while Cox agreed to postpone action on them for an indefinite period. In return for this significant concession, Cox agreed to give due consideration to Shaikh

Abdullah's request for supply of arms.[79] In January 1917, Cox was authorized to supply Shaikh Abdullah with a gift of 300 rifles and 100 rounds of ammunition per rifle.[80]

The outstanding issues having been settled, a final version of the treaty was produced. The historic Anglo-Qatari treaty was signed on 3 November 1916, by Shaikh Abdullah and Cox. Article III, relating to the arms trade was more stringent than the original draft while, as a concession to Shaikh Abdullah, article VI was altered to apply to British merchandise only and not to British traders. Articles VII, VIII and IX of the agreement corresponded respectively to articles IV and VIII of the draft. These three articles were left as they stood, but in a letter from Cox to Shaikh Abdullah of the same date as the treaty, it was agreed that they should be suspended for an unspecified period. It was agreed that, notwithstanding article IX of the treaty of 1820, with the Trucial chiefs, to which he had not subscribed, Shaikh Abdullah and his subjects should be allowed to retain slaves already in their possession.[81] In line with article III, Shaikh Abdullah signed the draft proclamation on arms traffic and issued it on the same date.[82] The treaty was ratified by the Viceroy and Governor General in Council of India on 23 March 1918.[83]

CONCLUSION

Although the 1916 treaty was the natural outcome of the laborious Anglo-Ottoman Convention of 1913, the flight of the Ottomans from al-Bida largely contributed for the conclusion of the treaty. British interest in the pearl fishing and the increased maritime warfare off the coast of Qatar were also contributory factors for the conclusion of the treaty.[84] However, it took more than two decades to conclude the treaty, placing Qatar under the British Trucial system of administration as the ninth and last of the Trucial states. The Qatar treaty went beyond the normal type of Trucial treaty, as it contained an understanding that the good offices of the British government would be granted to Shaikh Abdullah in the event of unprovoked aggression against him by land within the territories of Qatar. Shaikh Abdullah's prestige and image were certainly enhanced by conclusion of the treaty; he proved a tough negotiator. Through skilful negotiating tactics Shaikh Abdullah was able to keep the three offending clauses in abeyance for an indefinite period. Britain internationalised the issue of Qatar bringing it into the context of the Baghdad railway project and using it as a bargaining chip to gain concessions from the Ottomans on other fronts like Kuwait, Bahrain and Zakhnuniyah.

However, Shaikh Abdullah's position as the most important ruler in the region was recognised by Britain after the First World War, when he received the CIE on 3 June 1919, along with a 'salute of seven guns – honours of which he is the only recipients among the Chiefs of the Trucial Coast'.[85]

Notes.

1.   J. C. Hurewitz, Diplomacy in the Near and Middle East: A Documentary Record 1535-1956, vol. I, Gerrards Cross (U. K.): Archive Editions, 1987, p. 252.

2.   J. A. Saldanha, op. cit., Précis of Kuwait Affairs, 1896-1904, vol. V; see also Lorimer, op. cit., Historical, vol. IB, pp. 1026-7.

3.   Lorimer, op. cit., Historical, vol. IB, p. 1027; see also Busch, op. cit., pp. 193-4; and Memorandum by Harcourt, 'Respecting Kuwait' no. 7696, 29 October 1901, L/P&S/18/B133.

4.   Hurewitz, op. cit., p. 252.

5.   For full text see Ibid., p. 265.

6.   H. Rahman, The Making of the Gulf War: Origins of Kuwait's Long-Standing Territorial Dispute with Iraq, Reading (U. K.): Ithaca, 1997, pp. 37-38.

7.   Mallet to the US/S, IO, no. 16 January 1911, L/P&S/10/162; see also The Persian Gulf Historical Summaries: 1907-1928, vol. 1, Gerrards Cross, (U. K.): Archive Editions 1987, p. 2.

8   Busch, op. cit., p. 322.

9   Ibid., p. 324.

10.   Memorandum by the FO, Appendix no. 1, 30 March 1911 in 'Report and Proceedings of the Standing Sub-Committee of

the Committee of Imperial Defence on the Persian Gulf' (CID Report) 1 November 1911, CAB 16/15.

11. Minutes of the Second Meeting, CID Report, 15 June 1911, CAB 16/15.

12. See the Conclusion of the Meeting, CID Report, 14 July 1911, CAB 16/15.

13. Memorandum communicated to the Ottoman Ambassador in London, no. 1, 29 July 1911, R/15/5/65.

14. Talal Toufiq Farah, Protection and Politics in Bahrain; 1896-1915, Lebanon: American University of Beirut, 1985, p. 191.

15. Busch, op. cit., p. 330.

16. Ibid.

17. As quoted in Busch, op. cit., p. 330.

18. Ibid.

19. See the Aide-Memoire communicated to the FO by Tawfiq Pasha, no. 1, 15 April 1912, F.O. 371/1484.

20. A. Parker to Djavid Bey, Demi-official, unnumbered, 18 April 1912, F.O. 371/1484.

21. As quoted in Busch, op. cit., p. 331.

22. See the Joint Minute by Hirtzel and Alwyn Parker, Assistant Clerk, FO, on the Ottoman Government's Memorandum communicated on 15 April 1912, no. 1, 24 April 1912, F.O. 371/1484.

23. Ibid.

24. Minute by Parker, no. 16000, 23 April 1912, F.O. 371/1484.

25. Busch, op. cit., p. 323.

26. As quoted in ibid.

27.   As quoted in ibid.

28.   Ibid., pp. 333-334.

29.   The IO to the FO, no. 1, 10 June 1912, F.O. 371/1485.

30.   The G of I to Crewe, enclosure in no. 1, 13 June 1912, F.O. 371/1485.

31.   Crewe to the G of I, enclosure in no. 1, 1 July 1912; IO to FO, no. 1, 1 July 1912; See also Minute by Parker, no. 26379, 24 June 1912, F.O. 371/1485.

32.   For the full text of the Memorandum, see Crewe to Tawfiq Pasha, no. 1, 18 July 1912, F.O. 371.1485.

33.   The S/S for India to the V in C, no. 1440, 7 April 1913, L/P&S/11/46.

34.   The V in C to the S/S for India, unnumbered, 14 April 1913, L/P&S/11/46.

35.   The IO to the FO, no. 1140, 17 April 1913, L/P&S/11/46.

36.   Busch, op. cit., p. 337.

37.   See Shakespear to Cox, no. 13-G, 28 May 1913, R/15/1/65.

38.   Cox to the SGI/FD, Simla, no. 1963, 22 June 1913, R/15/2/30.

39.   For full text, see the 'Convention Between the United Kingdom and Turkey Respecting the Persian Gulf and Adjacent Territories [with Maps]', no. 10515, London: Her Majesty's Stationary Office, 29 July 1913.

40.   See Article II of the convention in Ibid; see also J. B. Kelly, Eastern Arabian Frontiers, London: Faber & Faber, 1964, pp.107-108.

41.   Crow to Marling, Baghdad, no. 37, 14 June 1913, L/P&S/10/384.

42.   Trevor to Cox, no. 441, 22 July 1913, R/15/2/30.

43.   Trevor to Cox, no. 403, 3 July 1913, R/15/2/30.

44.   Cox to Foreign Department, Simla, no. 1242, 11 July 1913, R/15/2/30.

45.   Trevor to Cox, no. 448, 26 July 1913, R/15/2/26.

46.   The V in C to the S/S for India, no. 3233, 2 August 1913, L/P&S/10/384.

47.   The V in C to the S/S for India, no. 3234, 10 August 1913, L/P&S/10/384.

48.   SGI/FD to the Political Resident in the Gulf, no. S-302, 13 August 1913, L/P&S/10/386.

49.   The V in C to the S/S for India, no. P-3284, 13 August 1913, L/P&S/10/386; see also Minute by the IO, no. 3914, 4 September 1913, ibid.

50.   Cox to the SGI/FD, no. 5002, 23 August 1913, L/P&S/10/386; see also V in C to the S/S for India, no. 3749, 11 September 1913, ibid.

51.   Ibid.

52.   Cox to Amir Abdul Aziz bin Abdur Rahman bin Faisal al Saud, no. 338, 11 September 1913, L/P&S/10/384.

53.   For al-Saud's reply, see V in C to the S/S for India, no. 4414, 20 October 1913, L/P&S/10/384.

54.   Ralph Paget (FO) to the US/S for India, no. 49302/13, 31 October 1913, L/P&S/10/384.

55.   Grey to Hakki Pasha, no. 29003, 1 July 1914, F. O. 372/573.

56.   Knox to Cox, no. 203, 1 September 1914, L/P&S/10/386.

57.   Knox to Cox, no. 204, 1 September 1914, L/P&S/10/386.

58. Ibid.

59. Knox to Cox, no. 238, 15 September 1914, L/P&S/10/386.

60. See the FO to the IO, no. 62022/14, 27 October 1914, L/P&S/10/386; see also T. W. Holderness (IO) to the FO, no. 2182, 21 October 1914, ibid.

61. The S/S for India to the V in C, no. 1502, 1 March 1915, L/P&S/10/386.

62. The Deputy SGI/FD to Cox, no. 40E-A, 13 May 1915, L/P&S/10/386.

63. Ibid.

64. Cox to the SGI/FD, no. B, 11 August 1915, L/P&S/10/386.

65. 'El-Katr: Turkish Evacuation' in the Senior Naval Officer, Bushire to the SGI/FD, no. P, 22 August 1915, ADM137/1152; for more details on the evacuation proceeding, see the Appendix no. 8.

66. Cox to the SGI/FD, no. 17327, 28 August 1915, L/P&S/10/386.

67. For the full text of the Draft Agreement between the British government and Shaikh Abdullah bin Jasim al-Thani, see Knox to the FS to the G of I, no. CG-52, 21 July 1914, R/15/2/30.

68. Cox to the SGI/FD, no. 17327, 28 August 1915, L/P&S/10/386.

69. The V in C to the US/S, IO, no. 3296, 8 September 1915, L/P&S/10/386; see also the SGI/FD, no. 900-S, 2 September 1915, ibid.

70. The S/S for India to the V in C, no. 3614, 9 September 1915, L/P&S/10/386.

71.   Cox to Grey, unnumbered, 16 September 1915, L/P&S/10/386; see also Cox to Shaikh Abdullah bin Jasim al-Thani, unnumbered, undated, ibid.

72.   For the Anglo-Qatari Treaty negotiations see Grey (from 'Lawrence' at sea) to Cox, unnumbered, 26 October 1915, L/P&S/10/386.

73.   Ibid.

74.   Ibid.

75.   Ibid.

76.   Cox to the SGI/FD, no. 3206/9-3, 17 April 1916, L/P&S/10/386.

77.   The V in C to the S/S for India, no. 2147, 2 June 1916, L/P&S/10/386.

78.   See the FO to the IO, no. 120978/16, 25 June 1916, L/P&S/10/384; see also Minute by Hertzel, IO, no. 2147, 22 June 1916, L/P&S/10/386.

79.   For the detailed account of Shaikh Abdullah-Cox negotiations, see Cox (HMS 'Lawrence' at sea) to the SGI/FD, no. T-15, 4 November 1916, L/P&S/10/386.

80.   The V in C to the SU/S, IO, no. 95, 10 January 1917, L/P&S/10/386.

81.   For full text of the Treaty, see Appendix no. 9. Contrasts with Busch as he stated no suspension of the articles VII, VIII & IX. In contrast of the article X, Busch also stated protection 'by land' instead of protection 'by sea' see Busch, op. cit., p. 146.

82.   See "Appendix C", proclamation by Shaikh Abdullah regarding the arms traffic, no. 2(b), 3 November 1916, L/P&S/20/C/158E.

83.   The Deputy SGI/FD, to J. H. H. Bill, Deputy Political Resident in the Gulf, no. 219-EA, 24 April 1918, L/P&S/10/386.

84.   Yousuf Ibrahim al-Abdullah, op. cit., pp.30-31.

85.   'El-KATAR 1908-16', Memorandum by India Office, no. B. 402, 5 September 1928, L/P&S/18/B.301.

CHAPTER TEN

Summary and Concluding Remarks

The process leading to the independence of Qatar was completed with the signing of the long awaited Treaty of 1916; the historical foundations upon which this treaty rested were Britain's 1853 Treaty with the Trucial Shaikhs, the treaty of 1861 with Bahrain as well as the Qatar Treaty of 1868. Article eleven of the unratified Anglo-Ottoman convention of 1913, recognised the independence of Qatar, under the leadership of Shaikh Jasim and his heirs, but Britain was unable to achieve its objectives in Qatar immediately, because of the death of Shaikh Jasim, initiator of the treaty, and the outbreak of the First World War. Britain had to wait until the Ottoman troops themselves deserted al-Bida fort, the symbol of the Ottoman empire in Qatar, which cleared the way for the opening of formal Anglo-Qatari negotiations. Unlike the Anglo-Qatari Treaty of 1868, the Treaty of 1916 was freely negotiated. It differed from the British treaties with the Trucial states in so far as the British agreed to grant Shaikh Abdullah good offices against any attack by land and to suspend three unwelcome clauses for an indefinite period. Shaikh Abdullah's extraction of these concessions from Britain was a diplomatic victory. While Britain achieved its long cherished aim of integrating Qatar into the chain of communications linking Britain and India and establishing the Trucial system in Qatar, Shaikh Abdullah gained some degree of protection from external threats to Qatar by land.

Shaikh Jasim emerged as the strongman of Qatar as a result of his successful uprising against the Ottoman military occupation of al-Bida; he first requested a British protection treaty in 1893 verbally to Resident Talbot. He made the same request following the British bombardment of Zubara harbour, despite the harsh and unjust treatment suffered by him and his allies at British hands. Shaikh Jasim's unhappy fate on this occasion was due to Ottoman betrayal and ineffectiveness in the Zubara crisis, which stood in stark contrast to the rhetoric and promise of support by Ottoman official in the run-up to the British attack. In fact, it seems possible that the Ottomans, eager to avenge Shaikh Jasim's victory over their forces at the battle of Wajbah, an event burned into the Ottoman memory, masterminded the Zubara crisis in order to destroy him.

Unwilling to agree to Shaikh Jasim's plea for British protection treaty, the British stationed a political agent at Bahrain to counterbalance the growing Ottoman influence on the Arabian coast and scrutinize the affairs of Qatar. At the beginning of the twentieth century, Shaikh Ahmed, Shaikh Jasim's deputy, who had been liaising with the Ottomans at al-Bida, made the case for a protection treaty with renewed vigour; he made several attempts to drive the Ottomans from al-Bida until his death in 1905. However, the views of the Foreign Office backed by the British Ambassador at Constantinople, were opposed to those of the India Office, which had strong support from the British authorities in the Gulf as well as from the government of India. This opposition from the Foreign Office demolished the chance of concluding a treaty with Qatar, despite the urgent regional need for one in view of the country's emergence as an entrepot for the arms trade.

The Ottoman occupation of Zakhnuniyah island changed the course of Anglo-Ottoman relations in the Gulf and emphasised the importance of a treaty with Qatar. While putting diplomatic pressures on the Porte for an early withdrawal from the island, Britain quickly decided to participate in the Baghdad-Gulf railway project as an equal partner and took steps to resolve the Qatar treaty issue within the broader framework of the Anglo-Ottoman disputes in the region as a whole. Despite ups and downs in the negotiations and differences among members of the British negotiating team themselves, the Ottoman and British negotiators were eventually able to narrow down their differences and reach an understanding on all the thorny issues in the Gulf. In this connection it is interesting to note that when Qatar emerged as a contentious issue in the negotiations, Parker, the British Foreign Office's representative urged his colleagues to surrender Qatar and recognise Ottoman suzerainty over it in exchange for Ottoman endorsement of the secret Anglo–Kuwaiti Treaty of 1899. Parker did this because of his determination to preserve British interests in

Kuwait, and a belief that failure to secure a treaty with the Ottomans would drive them towards the German camp. Nevertheless, contrary to Parker's hopes, the Ottomans joined with the Germans against the allies soon after the war broke out in Europe, despite the signing of the convention of 1913. The British ended up using Zakhnuniyah island, but not Qatar, as a diplomatic pawn to get the convention signed. Ottoman recognition of the independence of Qatar was a solid gain for Britain, although failure to define precisely the boundary of Qatar left the door open for future border disputes. The Ottomans thus abandoned Qatar in return for British recognition of their suzerainty over Zakhnuniyah, including the strip of coast between the island and Oqair; the Ottomans were desperate to gain control over the latter.

While the Anglo-Ottoman convention of 1913 established the independence of Qatar and ended more than four decades of Ottoman military occupation of al-Bida, Qatar had already emerged as a separate entity in 1868, as a result of foreign and regional powers' rivalries as well as the destruction of coastal settlements on the west and north coasts of the peninsula. The Portuguese destruction of the coastal villages was a key event for Qatar, spurring the growth of new settlements on the western coast of Qatar particularly at Furaiha which was later merged with Zubara. While harbour of Zubara existed since time immemorial, the al-Khalifa branch of the Utub accelerated the development of Zubara and they built their own fort called 'Qalat al Murair' two years after their arrival there; this they did in order to establish their hegemony in and around Zubara and drive out the original settlers. They soon established their own administration and monopolised the trade and commerce of Zubara, adopting a free trade policy. However, Zubara's rapid emergence as one of the Gulf's most thriving ports increased economic and political rivalries, which led to Persian Arab raids and attacks in an attempt to wreck the town's prosperity. Persian designs on Zubara came to an end once and for all with the Zubarans conquest in 1783, of the strategic Persian island of Bahrain, a campaign in which all the leading Qatari tribes and the Kuwaiti Utub participated. Nevertheless, the conquerors soon divided into two opposing groups because of the al-Khalifa's refusal to share the war booty and administrative power with one of their strongest allies; the Jalahima section of the Utub headed by Rahma bin Jaber, who mounted numerous attacks against al-Khalifa shipping and trade until his death in 1826.

Zubara continued to enjoy the status of a free port and substantial trade with India, central Arabia, Kuwait, Basra and Aleppo, despite al-Khalifa's transfer of their seat of government and financial activities to Bahrain. The decline of Zubara began with the establishment of Wahabi rule over the entire western coast of Qatar, with the strong support of bin Jaber. The Wahabis occupied Zubara to turn it into a strong political and military base; it also became home to anti-Bahraini forces as well as refugees dislodged from Muscat. Eventually, heavy Wahabi involvement in regional power politics and dynastic rivalries in Muscat led the latter to destroy Zubara (the Wahabi power base) and Khor Hassan in 1811 creating chaos and confusion along the entire north-west coast of Qatar.

The settlers gradually moved to safer places such as al-Bida, which was unexplored and unnoticed by any geographers or any senior British officials until the East India Company's warship *Vestal* bombarded it in 1821, forcing the inhabitants to flee. Despite the bombardment, al-Bida continued to grow in importance; it gained prominence when Macleod made the first visit by a British Political Resident to the coast of Qatar. However, the geography of Qatar along with its resources, towns, villages and tribes as well as the system of government, was first revealed to the outside world by the British survey of the Qatar coast, first published in 1829. While al-Bida emerged as the most important town in the 1820s, the country was not unified under one government or leader and the tribal system of government remained characteristic. In the absence of central authority, regional forces like the al-Khalifa and Wahabis, as well as the British empire, became involved in the affairs of the peninsula, which severely affected the society, economy and tribal unity of the country. The struggle for power in Bahrain dragged in the coastal towns of Qatar and had wide repercussions for the country as a whole as dethroned rulers and dissidents took shelter there and attempt to regain power in Bahrain with the support of the Qataris. Whoever held power in Bahrain felt uncomfortable unless places like al-Huwailah, Fuwairet and al-Bida were brought under his control either by force or applying the policy perfected by the British empire, of divide and rule. Bahrain's destruction of al-Bida in 1827 and Shaikh Abdullah of Bahrain's actions in al-Huwailah, which forced its chief bin Tarif to go into in exile in 1835, are prime examples in this regard. In pursuing control of the vital Qatar coast, Bahrain resorted to intimidation and interference, which led to the battle of Fuwairet in 1847 between the Bahraini chief Shaikh Mohammad

238

and bin Tarif, who returned to al-Bida in 1843, ending his long exile. Although bin Tarif was defeated in the battle of Fuwairet, it nonetheless sowed the seeds of the independence of Qatar and set an example of Qatari prowess in battle.

The battle of Fuwairet, despite the tragic loss of lives and the destruction it caused, contributed to the rise of Shaikh Mohammad bin Thani as a key figure in the history of Qatar. While retaining his chieftaincy of Fuwairet, in 1850 he moved to Doha, of which he also became the chief; Doha formally became the capital of Qatar in 1916. Contrary to Lorimer's remark that 'nothing is known of the manner in which the Al-Thani had attained by 1868, to predominant influence in Qatar',[1] it became clear that bin Thani rose to prominence and gained power as the result of the battle of Fuwairet and the brief Wahabi occupation of Doha and al-Bida, laying the foundations of the al-Thani dynasty in Qatar.

The main aim of the Wahabi chief Faisal's occupation of the eastern coast of Qatar was to invade Bahrain with the help of the Qataris and Bahraini dissidents because of Bahrain's failure to pay Zakat. While the Wahabi chief soon left Qatar, having concluded a peace agreement with Bahrain allowing Shaikh Ali to return to al-Bida, bin Thani was able to strengthen his position by forming a confederation with the leading tribes, the al Bu-Kawara, al Bin-Ali, the Sudan and al Bu-Ainain to safeguard the peninsula from Bahrain's interferences and raids. Doha became the headquarters of this newly formed confederation.

The emergence of Doha confederation consisting of Wakra, Doha and al-Bida as well as Fuwairet, under the leadership of bin Thani during the period from 1851 and 1866, was a severe blow to Bahrain's ambition in Qatar. Relations between the confederation and Bahrain turned from bad to worse when the latter's representative, stationed at Wakra, was expelled by the confederated powers in June 1867, because of his seizure and deportation to Bahrain of a Bedouin of Qatar without valid reason. Hostilities mounted between the two powers over this issue and culminated in the joint Bahrain-Abu Dhabi destruction of the towns of Wakra, Doha and al-Bida and in Qatar's abortive expedition against Bahrain in June 1868. The British Resident Pelly, who had been watching the situation closely, quickly intervened. Deposing Shaikh Mohammad bin Khalifa for violating the maritime peace and replacing him with his brother Shaikh Isa bin Ali, Pelly signed the 1868 Treaty with bin Thani at Wakra. Bin Thani's agreement to abstain from maritime warfare and refer disputes to the British Resident, and Pelly's address to all the leading tribal leaders of Qatar, signified that the Resident had established some influence over the tribes, extending the limit of the Doha confederation, and recognising the separate entity of Qatar for the first time. The 1868 Treaty, which effectively completed the unification of Qatar under one leadership, was the first step in the emergence of Qatar as an independent country, while the 1916 Treaty was the logical conclusion of that process. Shaikh Mohammad bin Thani was the founder of al-Thani dynasty, Shaikh Jasim bin Thani was the pioneer of the 1916 Treaty and Shaikh Abdullah bin Jasim al-Thani the torch-bearer of that dynasty.

Note

1.   Lorimer, op. cit., Historical, Vol. IB, p. 802. Lorimer's incomplete assessment on the situation in Qatar and his failure to narrate the circumstances under which Shaikh Mohammad bin Thani emerged as the most prominent figure of Qatar because of his reliance on the Political Agent Prideaux's report of 1905 on Qatar, see Prideaux to Cox, no. 208, 28 June 1905, R/15/2/26.

APPENDIX – I

Report on the Battle of Fuwairat, 1847

Translation of a letter from Hajee Jassem, Agent at Bahrein dated 12th Zilhuj / 21st November 1847.

A.C.

I beg to report intelligence of the Sheiks. It pleased God and they encountered on the 8th Zilhuj on which day Sheik Esai ben Tareef was killed and the people of Biddah [Bida] were completely defeated and of them were slain to the number of 80 men including ten persons of note of the family of Al Ali, of the Tribe of Esai ben Tareef, of the forces of the family of Khuleefa were killed 10 men. The combat lasted half the day and very many of the people of Biddah, were wounded whom the Sheiks (of Bahrein) embarked in two vessels and sent to Biddah and wrote letters to the Biddah authorities saying "God has decided matters between us, and Esai ben Tareef has been killed as you have seen. Should any be for us and desire to save themselves let them approach us and they and theirs are safe without question except the family of Al Ali, of the tribe of Esai they are not included in this amnesty" and they sent these communications to the people of Biddah and so in like manner they sent letters of rejoicing by a Buggarah to Bahrein and on the receipt of the intelligence the inhabitants were glad and Guns were fired and the country was tranquilized. Busheer ben Ramah has now sent messengers to Fysul and letters announcing the glad tidings to Lahsah [al-Hasa] and Kuteef [Qatif] of the occurrence of events and of the victory of the Sheiks. The country is now quiet and the inhabitants are relieved of their fears - they have now only to learn from Biddah what will be their answer - whether they will resign their authority and country into the hands of Sheik Mahomed ben Khuleefa or refuse to do so. Should they tender submission there will then be no further disturbance and apparently they have no other resource but to yield as there was no one (of consequence) amongst them but Sheik Esai and he is dead. Busheer ben Ramah has written to the Authorities of Kuteef and Lahsah not to send any forces whatever not even one man and should troops have been assembled to disband them as the affair was settled and it remained but to await intelligence from Biddah - that should that too be satisfactory all operations would cease.   True translation
/Signed/.
A. B. Kemball
Assistant Resident etc.

APPENDIX – II

Col. Pelly's Findings on Zakat Paid by the Chiefs of Qatar and Bahrain to the Wahabi Amir

No. 127 of 1866
Bushire 18th December 1866.

To,

The Secretary to Govt.
Bombay.

Sir,

I have the honor to acknowledge the Govt. letter No 3573 of the 24th ultimo requiring information as to what extent Bahrein is held to be actually subordinate to the Wahabee Rulers either in respect of tribute or otherwise.

2. The Islands of Bahrein are I believe held by the Sheikhs of Bahrein to be independent.

3. The Islands of Bahrein are held by English Authorities to be Independent and this view of the case has for at least the past 20 years been maintained by us. In exemplification of this view I would refer to correspondence now noted where [here] with reference (Letter from Major S. Hennell Rest. P. G. to Secy. to Govt. Bombay, d/ 10-5-1847 No. 3 Secret Dept. Govt. reply no. 334, Secret Dept. d/ 31st July 1847 & Govt. letter No. 512, Secret Dept. d/ 25th Nov. 1847 enclosing extracts Para 1 from the Hon. Secret Committee d/ 6th October 1847) to the designs of Turkey to establish a Supremacy over Bahrein Govt. laid down that it is obviously desirable to exclude as much as possible all interference by Foreign Powers in the affairs of the Persian Gulf since it is only by retaining the Supreme Authority in its own hands that the British Govt. can hope to secure the permanence of the objects it has gained in that quarter at such a large expense and where the Hon. the Board of Directors ruled that any attempts upon Bahrein ought to be resisted by the Naval Force in the Gulf and authorized Govt. to instruct the Resident accordingly. On a subsequent occasion it was laid down by the Govt. of India that H. M's Govt. would not admit of the occupation of Bahrein (vide Mr. Secy. Goldsmid's letter No. 166 Secret Dept. d/ 31-5-1853 to Capt. A. B. Kemball Rest. P. G. & enclo. from Govt. of India No. 177 d/ 10-5-53) by the Turkish Govt. or by any one acting for it or in its interest

and consequently that the G. of I. [Govt. of India] should offer every obstacle to an attack upon that Island by the Wahabee Chief - - Again Sir Charles Wood in his letter now noted (No. 2 Secret Dept. India Office Feb. 18th 1861) and copy of which has already been submitted decided that Bahrein should be regarded as Independent and subject neither to Turkey nor to Persia.

3.	Since writing the foregoing an Envoy from the Bahrein Chief, has called on me, acknowledging and apologizing for past misconduct and giving solemn assurances that for the future neither the English Govt. nor its subjects shall have any cause of complaint - - The envoy states that the Sheikh of Bahrein hold himself absolutely independent of the Wahabee Ruler in so far as the Islands of Bahrein are concerned - But the Chief having occupied certain lands in Guttur [Qatar] in the mainland of Arabia receives revenue from these lands and pays to the Wahabee Ameer the sum of (4,000) Four thousand Dollars per annum, on condition that the latter prevent his tribes from molesting the people residing on the Guttur lands.

4.	The Wahabee Ameer as I have already reported verbally (vide my report No. 57 of 15th May 1866 Pol. Dept. Para 74) informed me that God had given him all the land of Arabia from Koweit round to Rasul Hudd thus including not Bahrein only but the Muskat Territories and the Turkish Port of Koweit - These pretensions however would not I presume be allowed by Government.

5.	As a rule the Chief of Bahrein hoists his own flag but he is an eccentric man and has been known when a Turkish Official arrived to hoist with his own the English and Persian flags - On one occasion he tendered his Island to the English in gift. (Vide Col. Hennell's letter to Mr. Chief Secy. Malet No. 66, Secret Dept. d/ 28th Feb. 1849). But these proceedings have been due to the irritable temper of the Sheikh himself and this irritability in turn is said to be due to an over indulgence of his sexual passion.

6.	The records of this office contain very frequent complaints against the Bahrein Chief. As examples I beg to append two letters, one from the Wahabee Ameer, the other from my predecessor.

7.	For some months past the Sheikh has given no trouble and I am desirous of hoping that the overtures now made by him are sincere.

8.	For myself I can only assure Government that my persistent wish has been to conduct affairs with Bahrein as with all other Chiefdoms in a friendly manner and in literal conformity with the instructions of Government.

> I have etc.
> /Sd/ L. Pelly
> Pol: Rest. P. G.

For enclo. 1-vide Bahrein – subject   8 – only letter.
 "   "   2-vide Bahrein -   "   9 – first letter.

1. Sheikh Ali bin Khaleefa received one year's tribute from Katr tribes – How much?

1. 9,000 Krans of which 4,000 Krans were paid to the Shaikh of the Noaim Tribe in Guttur, and 5,000 Krans to the Chief of Bahrein direct.

2. Was this made over to the Wahhabees?

2. No, Shaikh Ali declined to pay pending the Imaum's declaration of good will and friendship which, his harbouring Naser bin Mobarek and late requisition on the Guttur Chief for a force of 30 Horsemen and 50 Camel-men, tended to make doubtful-vide Major Way's report dated 21st August 1869.

3. Sheikh Esau also received on year's tribute – Was this paid to the Wahhabee?

3. No – For similar reasons as above and Shaikh Esau wrote as follows:-
In regard to the 5,000 Krans belonging to the Wahabee Chief, I have not heard anything from the Wahabees. Abdallah bin Faysal is my enemy and does not disguise this. I cannot therefore give him the 5,000 Krans unless he be friendly to me. – vide letter dated 27th Rajab 1287 – 22nd October 1879.

4. How much did the Wahhabee claim from Bahrein and Katr?

4. 4,000 Dollars or 20,000 Krans.

5. Did Chiefs of Bahrein levy in late times or any time levy taxes for their own benefit from Guttur?

5. No tax whatever – The 9,000 Krans were considered as contribution by Guttur towards a total sum (viz:- 20,000 Krans) payable by Bahrein and Guttur combined in view to securing their frontiers from molestation by the Noaim and Wahabee Bedouins.

In a report to Govt. under date 18th Decr. 1866, Col. Pelly wrote to Govt. as follows:-

245

An Envoy from the Bahrein Chief
has called on me. x x x x x x x x
The Envoy states that the Sheikh of
Bahrein holds himself absolutely
independent of the Wahabee Ruler in
so far as the islands of Bahrein are
concerned. But the Chief having
occupied certain lands in Guttur on
the mainland of Arabia receives
revenue from these lands and pays to
the Wahabee Ameer the sum of 4,000
Dollars per annum, on condition that
the latter prevent his tribes from
molesting the people residing on the
Guttur lands.

With reference to the above – Govt.
of India wrote as follows:-
x x x x x x x x x x x x x x x x x x x x
regarding the position of the Chief
of Bahrein – I am directed to observe
that this Ruler would appear from
Col. Pelly's statement to be liable to
the Wahabee Chief for tribute on
account of certain lands held on the
Arabian Coast, and so far as such
territory is concerned it may be
presumed that he owes fealty to the
Wahabee Government. But in so far
as Bahrein proper comes in question,
H. E. in Council concurs in the
opinion expressed in your 2nd Para:
that the Chief is independent and
owes allegiance to no other power.

APPENDIX – III

Shaikh Mohammad bin Thani's acceptance
of the Ottoman Flag

Dated 3rd Jemadee-oos-Sanee 1288, received 14th August 1871.

Translated purport of a letter from MAHOMED BIN SANEE [Thani], Chief of Guttur [Qatar], to HAJEE ABDOOL NUBEE.

I beg to inform you that some days ago Abdullah ben Subah, Chief of Koweit, arrived at Guttur with a deputy from Fereck Pacha; they ordered me to establish a Turkish flag and I was unable to refuse hoisting the flag owing to their having supremacy on the land. You know also that we belong to the land, and are unable to refuse to obey their orders.

I hear now from them that they have seized Lahsa [al-Hasa] and those parts. God knows what may hereafter happen.

Major Smith, Assistant Resident, arrived here, and remained one day. Did not land and gave me no reply. I did not ascertain what object he had. He left for Oman.

I enclose a letter for the Resident, and beg you will deliver it and explain our condition to him, in that we are willing to have peace at sea and obedient to him and will not help towards disturbing the peace of the sea. But in regard to establishing the Turkish flag in Guttur, we belong to the land, and this Government have the supremacy on land so that we cannot refuse to obey their orders. I asked Major Smith to get our claims from the Ameer [Amaeer tribe], but he gave me no (satisfactory) reply, saying that this question must remain pending till the Ruler in Nejd is established.

You know that the Chiefs of the Ameer [Amaer] are independent. You are my Agent in all matters, and I beg you will explain my status to the Resident in such manner as you may think advisable.

APPENDIX – IV

Shaikh Jasim's Statement on Battle of Wajbah

Dated the (15th Shawal 1310) 2nd May 1893.

From   -   JASIM-BIN-MUHAMMAD-BIN-THANI.
To   -   LIEUT. COLONEL A. C. TALBOT, C.I.E. Political
               Resident in the Persian Gulf.

I beg to represent to you that I endeavoured to the utmost of my power to advise, serve, and protect the Government of which I am a subject, and none of its officers experienced the slightest opposition from me. In former days both I and my father were amongst the Shaikhs of the littoral, having treaties with you; but when the Turkish Government came towards Nejd I elected to give my allegiance to it, while other Shaikhs did not. I and my father were Walis of Katr [Qatar] and its dependencies; but on my becoming a subject of the Turkish Government, I was reduced to the place of a Kaimakam, and I remained in its service 24 years. Formerly I had only a small police force given me, and I protected it and the people of Katr. When the people of Hasa and others revolted and committed treason, I served with obedience and offered no opposition. I served gratuitously. Walis, Mutaserrifs, and Inspectors came to me, but they did not find in me the slightest fault; on the contrary, they returned with feelings of the deepest gratitude towards me.

When the present Wali, Hafiz Pasha, arrived he made a terrifying demonstration, by moving land forces both regular and of Police and irregulars, and ships of war by sea. He also asked the help of the Shaikhs of Koweit, of the Ajman and Al Murrah tribes, against me. When I saw that he was advancing by land in this manner and other forces were following him, and I found no reason for such an attitude, I beg to fear, as I had offered no opposition nor stirred rebellion. I became sure that his advance was not that of one coming to make peace and to allay excitement, but was aimed at destroying peace and raising disorder; for had his object been to ensure peace and quiet, he would have come in the manner others have come, seeing that I have been obedient to the Turkish Government and obeying its officers out of respect to it. When therefore I saw him come thus I went away from the country and out of his way, warning all my tribes not to interfere with him to his detriment, nor to create disturbances against him. I also warned the townspeople to go out to meet him so as to enable him to enter the town with the utmost honour and respect; I also told them that they should wait upon him and obey his orders.

Accordingly the tribes removed themselves from his way and abstained from interference. And the townspeople went out to receive him. The principal men waited upon him obediently and asked him for assurances of safety to allay their apprehensions. The "Aman", guarantee of safety in the name of God and then of the Government was accordingly proclaimed by a crier throughout the town, and the people became quiet and re-assured.

Then the Wali wrote to me for an interview. I replied that I was a servant of the Sublime Government and its officials, but had become apprehensive on account of the movement of the troops, and the assistance sought for from the Arab Shaikhs and tribes; that if he wanted Katr there it was, and its people, at his disposal – he could do what he liked and make any arrangements he chose; that I would not interfere in any affair calculated to oppose him; that if his object was to have an interview with me I could have had no excuse to avoid it, had his visit been like those of former Walis – I would have waited upon him before his inviting me, - but that he had frightened me by this military display, and under the circumstances it was not possible for me to meet him in the town; that if he wanted to see me he should come out of the town, alone, to some place, and I would then meet him.

When I replied in the above sense he asked to see my brother Ahmed. I asked "Aman" for him, which he gave in the name of God and the Turkish Government. Accordingly I sent my brother as my deputy, recommending him to obey orders. So when my brother arrived the Wali broke the promise, and thus brought into contempt the guarantee of the Turkish Government. He put a chain round the neck of my brother and ordered his imprisonment on board. He then seized twelve of the principal men of the town, put chains round their necks, tied their hands and feet, and subjected them to a rigorous and painful imprisonment on board. When I saw their fate and that of my brother my fear increased and the townspeople became terrified and perturbed at what had befallen their Chiefs. Then the Wali issued a notification to the obedient tribes of Katr, owing boats and property in the country, to the effect that they should stop all intercourse. This notification spread consternation among the people both by land and sea.

Struck with fear by the course of events, I wrote to him asking him for mercy, and representing that his action in imprisoning persons after he had given the guarantee of the Sultan was an act of treachery which is not customary with Turkish officers; that his imprisoning them would increase the fear of the people, who would disperse and the country be ruined in consequence.

I then entreated him to release the prisoners, promising to comply with any demands he might have on me in order to avert a general calamity resulting from the ruining of the country. He replied that I should surrender myself and disperse the tribes that were with me. I rejoined that I was willing and obedient, and that the tribes had not been assembled to fight him; for if I had done this I should have been fighting against the Turkish Government, in whose service I am, and to which I owe allegiance. Had I intended to fight him I would have done so before he entered Katr, and prevented him from entering it. That the tribes had assembled on perceiving that the Shaikh of Koweit-bin-Subah, with his tribes, was marching to Katr to attack them, and that they had therefore joined to protect their families and their honour. That if, however, their dispersion was necessary, I would endeavour to effect that, provided that he would release the prisoners. He replied that I should comply with his demands and send away the tribes, but the prisoners must be kept as hostages to provide against future disturbances. He called me a rebel, and said that I must come to sue for pardon with a rope round my neck.

When I saw that his reply was against all sense of justice and equity, I became aware that he wanted to create mischief. So I sent away the tribes and told them to go wherever they liked. I resolved to leave Katr altogether myself, taking my servants with me, and to complain against him in high places. When he learns that the tribes had dispersed, and that I had resolved to leave Katr, he collected troops, drew out the guns, and marched by night. I heard nothing of his movements until the morning of the 6th Ramazan when I was seven hours distance from the town, when I heard the sound of cannon and musketry against the houses, and there was so much of it that even some women and children were killed. Under these circumstances I was compelled to defend myself, being driven to extremity. When the sound of the cannonade roused the scattered tribes they advanced against him. The Wali then retreated and fled on horseback, leaving his troops behind him, after having raised this mischief and kindled the flames of war. The Imperial troops which he had hurled against the tribes retreated after him, and then there happened what has been witnessed. He reached the town at the close of the day, and could have protected the remaining troops and averted further mischief between them and the townspeople; but he did not remain firm – lost his presence of mind, and fled with Tahir Beg to the ship, leaving the troops without a commander, and then ordered the ship's people to fire on the townspeople, who, despairing of ever receiving justice, and being deprived of protection, and moreover, learning that the Wali was aiming at their destruction, fled and the town became deserted. The troops on shore seeing the ship firing, also commenced to fire at the people with muskets, and to plunder the town. Then a rising followed which is beyond all description. I heard this news when I was at my quarters in the interior; so I went with a short distance from the suburbs of the town, and sent persons to stop the people and the tribes from further fighting with the troops, and to remove them out of the town.

The people then came to me and asked that I should endeavour to have their Chiefs released, which I promised to do in order to ensure the safety of the remaining troops. The Wali delayed matters for three days, with the object of causing the destruction of the people and of the remaining troops. I therefore wrote to him informing him that if he did not release the prisoners great disaster would happen to all. He then preferred his demands, one of which was that I should arrange for the safe transit by land of El Hasa of the cavalry and irregulars; the other that I should restore to him in the ship his horse, treasure, and luggage, as well as other horses and things which he said were missing after the fight.

When he released the prisoners after having killed one, and put out the eyes of another, I complied with all his demands. I sent off to him the things he claimed, and also arranged for the same passage of the cavalry by land, by sending with them persons who would be responsible for their safety until they reached Hasa. I also arranged for the passage of those he wanted to send away by sea, by embarking them in boats.

The Wali remained on board the steamer facing the town, and threatened every passer by saying that he had asked the Turkish Government to despatch a man-of-war and large forces, and that his object was to destroy Katr and its people, and to raze the town to the ground.

When the people heard this they became reluctant to remain in the town, and my endeavours to dissuade them failed, and they all left to a man. The town became deserted to an extent never witnessed before. After this I went to the troops that were remaining in the town, and arranged for their protection. I kept some Arab watchmen to look after them and promised to give them anything they wanted for their sustenance until orders arrived from the Sublime Government. I remained in the mainland behind them to protect them against injury or molestation at the hands of any one. The people have dispersed, some amongst villages, and others to the suburbs of Katr.

I have now represented my case to the mercy of the Sublime Government, imploring it to consider my case with justice, as regards the loss and destruction of my property consequent upon the ruin of the town. I have lost what I possessed, and the little which the people owed me has been lost owing to their dispersion. I am now expecting to receive mercy from the Sublime Government. If I receive justice against Hafiz Pasha who began the quarrel, sacrificed the lives of the people and soldiers, destroyed property and country, scattered the subjects, and disturbed the tribes, well and good. But if the Sublime Government does not attend and repair the damage, then there is no doubt that the breach will be widened.

APPENDIX – V

Commander Pelly's Naval Action at Zubara

No. 98, dated Bushire, the 15th September 1895.

From - Colonel F. A. Wilson, Political Resident, Persian Gulf.

To - The Secretary to the Government of India, Foreign
      Department.

In continuation of my telegrams of the 12th instant, I have
the honour to report, for the information of His Excellency the
Viceroy and Governor-General in Council, that on the 6th instant
Commander Pelly, Senior Naval Officer, Persian Gulf Station,
having in his judgment, after careful consideration of the facts
immediately before him, determined that prompt action was the
only means of averting the plunder of Bahrein with its attendant
excesses, attacked and destroyed some 44 out of a fleet of native
craft assembled off Zobara, armed and ready for an instant descent
on that place. I enclose a copy of Commander Pelly's despatch of
the 7th instant and of its enclosures reporting the operations.

Commander Pelly's despatch must have been written
under pressure of haste and many demands on his attention, and it
seems desirable that I should supplement it with a statement of the
considerations bearing on the situation, within my own
knowledge.

2. Recently as has been reported, the chief representative
of Ottoman authority in this region, the Mutasarif of El-Hasa,
formally intimated that the tribes of Katr [Qatar] having passed out
of the control of authority, had determined on the attack of
Bahrein, which holds a number of British subjects, and a very large
amount of British-owned wealth. The Mutasarif accordingly,
repudiating all responsibility for the catastrophe which he thus
notified as imminent, called for the removal of all British subjects
within 17 days, an operation which would have been most ruinous
if not impossible. Term of grace fixed by the Mutasarif expired on
5th instant, and as will be seen from Commander Pelly's despatch,
the local representative of the Ottoman Government at Zobara then
boarded H. M. S. "Pigeon", which had been ordered there to watch
the hostile fleet of boats known to be there in pursuance of the

declared intention to attack Bahrein. This official ordered the departure of the ship enforced by a threat, reasserted the resolve to attack Bahrein, which had been notified by his superior the Mutasarif, and added that the Turks will join in this attack. The boats themselves were present, fully armed and ready to proceed at once upon the operation.

3. These deliberate utterances, and the facts, therefore, left no possible room for doubt, so far as words and acts can make an intention certain, that the attack would be delivered, and there can hardly be a question as to the nature of the excess and outrage which would attend an attack under such auspices. The only point on which a doubt could arise is whether this attack would be with the direct countenance and support of the Turks or not, and though the statement of Shaikh Jasim in his letter of the 7th that the hostile fleet had been collected by the order of the Mutasarif lends support to the word of the inferior officer (the Mudir), we are fully entitled to assume the full significance of the assurance of his superior, that the act would be that of irresponsible tribes beyond control, for which the Turkish Government could not answer. And here I would observe that the actual status of a person like Jasim, as to whether he is or is not a responsible officer of the Turkish Government, is most difficult to determine. On the one hand he is the declared Kaim Makam of Katr; on the other, the assertions of the Mutasarif would make him the virtual leader of tribes who are beyond the control of authority, and for whose acts no responsibility is accepted by the Porte. The dilemma is the creation of the Porte, whose full responsibility in regard to either view that may be adopted would seem absolute.

4. Commander Pelly, on arrival at Zobara, found himself called upon to deal instantly with the grave question as to the best means of repelling the imminently threatened attack on Bahrein. There seems nothing to prevent its delivery the same night, the moon being full, as had doubtless been considered in fixing the term of 17 days' grace. Had this opportunity been given, and taken advantage of, I conceive it would have been a matter of extreme difficulty and indeed probably impossibility, for the ships to have taken effective action against a crowd of boats. Commander Pelly has referred, in his despatch of 3rd, to the impossibility of following such a hostile fleet towards the south of Bahrein, by reason of the waters not being surveyed, and a reference to the chart would seem to indicate that the limits within which operations could be

carried on with due regard to the safety of H. M's ships, between Katr and the mainland generally are extremely limited. Any delay in acting would therefore seem to have involved the gravest danger of a disaster.

5. On the morning after the operations, the 7th September, Shaik Jasim wrote his first letter enclosed in Commander Pelly's despatch. It calls for little remark. I would only refer to the impudence of his reference to the appeal for 15 days' grace. The incident is dealt with in Commander Pelly's despatch of 23rd July (Enclosure in my letter no. 75 of 27th July), which also reports the detention, at the very time, of the boats sent from Bahrein for family and followers of Salim bin Hamad, who had made his peace with the Chief. To the significant assertion that the hostile boats had been collected under orders from the Mutasarif, I have already referred.

6. With regard to Commander Pelly's reply imposing certain terms as the conditions of immunity for the remainder of the boats at Zobara, the first is in accordance with the original orders of Government. I would venture, however, to ask a re-consideration of the point as to insistence on the return of the whole Al bin Ali tribe to Bahrein. Probably many, perhaps the greater part, will elect to do so, but the relations of some with Sheikh Esa may now be such as to cause a mutual desire for separation. I would ask whether in such case, evacuation of Zobara being rigidly enforced, the condition as to return to Bahrein might be relaxed. As to the fourth condition, the number of boats delivered over is not yet known, nor whether these belong, for the most part, to the Al bin Ali or to Jasim's people; the Residency Agent at Bahrein, however, reports that crews sufficient to navigate about 100 boats have been requisitioned.

Since writing the above I have received your telegram of this date.

I regret to add my fear that Commander Pelly has suffered very severely from the severity of the hot season of this climate, to the extreme of which he has been exposed for over two months. The heat is now abating.

APPENDIX – VI

Ottoman Intrigue Against Shaikh Jasim

From Viceroy, 14th September 1895.

Commander Pelly having sent "Pigeon" to watch hostile dhows at Zobara received report on 5th September that Mudir boarded "Pigeon" and ordered her immediate departure, using threats in case of refusal, and intimating that Jasim would attack Bahrein and Turkey would join in attack. Pelly proceeded to Zobara on 6th September, and finding boats armed ready for attack determined that destruction of hostile fleet was only way to prevent attack and plunder of Bahrein. Written warning was conveyed to Jasim in the afternoon, and after one hour, "Pigeon" and "Sphinx" opened fire and destroyed 44 dhows. On the morning of the 7th September Turkish officials and flags no longer at Zobara. Jasim flew flag of truce and wrote offering to surrender, begging for pardon and stating that the Mutasarriff had ordered the collection of boats. Terms offered by Pelly were:- Al-bin-Ali evacuating Zobara and returning to Bahrein, dispersion of Bedouin near Zobara, and restoration of the Bahrein boats. Jasim accepted terms, pleading no desire to oppose.

APPENDIX – VII

Shaikh Jasim's Message for British Protection of Ibn Saud

No. 500 of 1906
Political Agency Bahrain
Dated the 17th of November 1906

Confidential

To,

The Political Resident in the Persian Gulf,
Bushire.

Sir,

I have the honour to report that on the 10th October 1906 I received a brief note, translation attached, from Shaikh Jasim bin Thani, written with his own hand, requesting me to visit him for a few minutes' conversation.

2.   It was inconvenient for me to go to Lusail within a reasonable time after the receipt of this note, I therefore deputed my Interpreter, Mr. Inam ul Hak, to visit Shaikh Jasim, instructing him only to receive the old Chief's message, and not to give any reply in my name. Mr. Inam ul Hak reached Lusail on the 23rd October 1906, and left the following day for Doha in the boat by which he had travelled from Bahrain, as he would otherwise have been detained at Lusail for an indefinite time waiting for a means of return.

3.   Shaikh Jasim's message transpired to be connected with the affairs of Abdul Aziz bin Sa'ud. The latter had requested his friend to communicate with me, because he suspected that the Chief of Kowait had been representing his cause to the Government of India in lukewarm manner and it seemed possible that Shaikh Mubarak was afraid of his own importance being overshadowed if Abdul Aziz succeeded in achieving his ends.

4.   The resources of Najd are stated to have been strained to the utmost by the recent internecine wars, and Bin Sa'ud considers that the Oases of Hasa and Katif were always the most profitable possessions of his Wahhabi ancestors. He is anxious therefore to recover the two districts and he proposes that a secret

understanding should be arranged between the British Government and himself under which he should be granted British protection from Turkish assaults at sea, in the event of his ever succeeding in driving the Turks, unaided, out of his ancestral dominions. In return for this protection the Amir is willing to bind himself to certain agreements (probably similar to those of the Trucial Chiefs) and to accept a Political Officer to reside at his Court. The details of this secret treaty he wishes to be settled or discussed at an interview which he is ready to give me either in person, or with his brother representing him, at some convenient rendezvous in the desert. Bin Sa'ud is determined to make an effort to obtain possession of Hasa and Katif, for without the additional revenue which he can derive from these tracts, he admits that he is unable to control the tribes who menace the highways of commerce and pilgrimage. He proposes therefore, in the first place, to apply to the Sultan for the Mutasarrif-lik (Governorship) of the districts, and to throw off the Turkish yoke as soon as he considers the moment favourable, after establishing himself. If his application is refused, he will invade the districts as soon as he is ready, and having captured them, he will appeal openly to the British Government for protection. If he fails, he will never betray the secret understanding between himself and the Government! Possibly he will not make his attempt, even, for 4 or 5 years more.

5.   Such is the message, and Shaik Jasim requests that the reply may be given to him only by word of mouth, when he will arrange to forward it on by trusted messenger.

6.   I presume that the only answer possible is that while the British and Turkish Governments entertain friendly relations towards each other, it is impossible for the Government of India to enter into any such agreement, as the Amir desires. I would like, however, to be permitted to add a warning to the Amir, that if he has never yet involved himself in any direct admission of subordination to the Sultan, as I understand he claims not to have done, he should be careful not to commit himself now unless he decides to abide by the consequences. It is generally believed by the Arabs in Bahrain, that the Turkish Government, finding that they are unable to collect revenue in the two districts and that the people are favourable towards Bin Sa'ud, are themselves contemplating an invitation to the latter to assume the government in the Sultan's name.

7.   With regard to Shaikh Jasim's own health, my Interpreter reports that he appeared considerably stronger than last spring, and that his eyes are so much improved that he is able to attend to all his correspondence himself.

He seemed however to be downcast in mind, which Mr. Inamaul Hak, subsequently learnt was due to the fact that he had invested 80 lakhs of rupees in pearls this season at very high rates, and that he had found recently that he was likely to lose heavily in consequence of the subsequent fall in prices.

I have the honour to be,
Sir,
Your most obedient servant
(Sd.) F. B. Prideaux. Captain.
Political Agent, Bahrain.

APPENDIX – VIII

The Ottoman Evacuation of al-Bida Fort

H. M. S. "Pyramus" at Doha,
20th August, 1915.

Sir,

I have the honour to report as follows:-

H. M. S. "Pyramus" arrived at Bahrein at 8.40 a.m. 18th August 1915, and I saw Major Keyes, who informed me that the last Tangistani Dhow left Bahrein on 13th August.

The guns of H. M. S. "Juno" were distinctly heard on the Arabian coast during the attack on Dilwar, and all Tangistani Dhows went into hiding.

Major Keyes embarked on board "Pyramus" and Ship sailed from Bahrein at 2.2 p.m. 18th August, arriving 5 miles off Doha at 11.30 a.m. 19th August. It took till 2.15 p.m. to negotiate the entrance, the water being very shallow, and a boat having to be sent ahead. At 2.15 p.m. "Pyramus" and "Dalhousie" anchored just inside entrance of harbour. Examined several Dhows which were making out to sea. "Pyramus" then proceeded up harbour and anchored 2,000 yards from Turkish Fort, and got out stern anchor in order to keep broadside bearing. Strong entrenchments were observed between Fort and sea with field guns in them.

The Sheik of Qatar visited Ship and had a conference with Major Keyes, in which it was arranged that he should endeavour to induce the Turks to abandon the Fort, leaving the guns intact. He further informed Major Keyes on his word of honour, that no Tangistani Dhows were at Doha. (This was confirmed by enquiries made on shore).

The Sheik's visit was returned at 7.30 a.m. 20th August, and he informed Major Keyes that the Turks had evacuated the Fort during the night.

A party landed from H. M. S. "Pyramus" and took possession of Fort and entrenchments, finding 3 Field Guns with breech Blocks removed, 14 Rifles and 120 cases of ammunition, about 500 projectiles – most of them fuzed – tents and other Military stores. By Major Keyes advice the Rifles and ammunition were turned over to the Sheik of Qatar with some of the stores. I embarked the Guns on board "Pyramus" and threw overboard the projectiles. I considered these were dangerous as no one on board understood the fuzes, and, as no powder charges for the field guns were found, the projectiles would not be of much use for the Turkish Guns at Bushire.

I enclose sketch of Fort and entrenchments by Sub Lieutenant Eric R. A. Farquharson.

"Pyramus" left Doha at 3.20 p.m. on 20th August, and proceeded on a course for Bahrein.

On receiving your message proceeded to Ras-Naband, and searched all Dhows, greatest number being off Nakhil-Takki, no Tangistani Dhows being found.

Major Keyes landed at this village and questioned the inhabitants as to Tangistani Dhows, but they could give no information.

Proceeded to the Town of Kangun and on the way there, a thick sand storm and a strong N. wind came on. In view of these circumstances I decided not to call there owing to it being a lee shore and to the thick weather making navigation difficult.

Proceeded to Bushire.

> I have the honour to be,
> Sir,
> Your obedient Servant,
> Sd/- Kelliwin.
> Commander.

The Senior Naval Officer,
   Persian Gulf,
      H. M. S. 'Juno".

APPENDIX –IX

Treaty between the British Government and Shaikh Abdullah bin Jasim bin Thani, Shaikh of Qatar, dated the 3rd November 1916.

Whereas my grandfather, the late Shaikh Mohammad bin Thani, signed an agreement on the 12th September 1868 engaging not to commit any breach of the Maritime Peace, and whereas these obligations to the British Government have developed on me his successor in Qatar.

I.

I, Shaikh Abdullah bin Jasim bin Thani, undertake that I will, as do the friendly Arab Shaikhs of Abu Dhabi, Dibai, Sharjah, Ajman, Ras-ul-Khaima and Umm-al-Qawain, co-operate with the High British Government in the suppression of the slave trade and piracy and generally in the maintenance of the Maritime Peace.

To this end, Lieutenant Colonel Sir Percy Cox, Political Resident in the Persian Gulf, has favoured me with the Treaties and Engagements, entered into between the Shaikhs abovementioned and the High British Government, and I hereby declare that I will abide by the spirit and obligations of the aforesaid Treaties and Engagements.

II.

On the other hand, the British Government undertakes that I and my subjects and my and their vessels receive all the immunities, privileges and advantages that are conferred on the friendly Shaikhs, their subjects and their vessels. In token whereof, Sir Percy Cox has affixed his signature with the date thereof to each and every one of the aforesaid Treaties and Engagements in the copy granted to me and I have also affixed my signature and seal with the date thereof to each and every one of the aforesaid Treaties and Engagements, in two other printed copies of the same Treaties and Engagements, that it may not be hidden.

III.

And in particular, I, Shaikh Abdullah, have further published a proclamation forbidding the import and sale of arms into my territories and port of Qatar; and in consideration of the undertaking into which I now enter, the British Government on its part agrees to grant me facilities to purchase and import, from the Muscat Arms Warehouse or such other place as the British

Government may approve, for my personal use, and for the arming of my dependents, such arms and ammunition as I may reasonably need and apply for in such fashion as may be arranged hereafter through the Political Agent, Bahrein. I undertake absolutely that arms and ammunition thus supplied to me shall under no circumstances be re-exported from my territories or sold to the public, but shall be reserved solely for supplying the needs of my tribesmen and dependents whom I have to arm for maintenance of order in my territories and the protection of my Frontiers. In my opinion the amount of my yearly* requirements will be up to five hundred weapons.

IV.

I, Shaikh Abdullah, further undertake that I will not have relations nor correspond with, nor receive the agent of any other Power without the consent of the High British Government; neither will I, without such consent, cede to any other Power or its subjects, land either on lease, sale, transfer, gift, or in any other way whatsoever.

V.

I also declare that, without the consent of the High British Government, I will not grant pearl-fishery concession, or any other monopolies, concessions, or cable landing rights, to anyone whomsoever.

VI.

The Customs dues on the goods of British merchants imported to Qatar shall not exceed those levied from my own subjects on their goods and shall in no case exceed five per cent, *ad valorem*. British goods shall be liable to the payment of no other dues or taxes of any other kind whatsoever, beyond that already specified.

VII.

I, Shaikh Abdullah, further, in particular, undertake to allow British subjects to reside in Qatar for trade and to protect their lives and property.

VIII.

I also undertake to receive, should the British Government deem it advisable, an Agent from the British Government, who shall remain at Al Bidaa for the transaction of such business as the British Government, may have with me and to watch over the interests of British traders residing at my ports or visiting them upon their lawful occasions.

---

* Note. – In the original Treaty in the English version the word "early" has been written for 'yearly" by slip of the pen.

IX.

Further, I undertake to allow the establishment of a British Post Office and a Telegram installation anywhere in my territory whenever the British Government should hereafter desire them. I also undertake to protect them when established.

X.

On their part, the High British Government, in consideration of these Treaties and Engagements that I have entered into with them, undertake to protect me and my subjects and territory from all aggression by sea and to do their utmost to exact reparation for all injuries that I, or my subjects, may suffer when proceeding to sea upon our lawful occasions.

XI.

They also undertake to grant me good offices, should I or my subjects be assailed by land within the territories of Qatar. It is, however, thoroughly understood that this obligation rests upon the British Government only in the event of such aggression whether by land or sea, being unprovoked by any act of aggression on the part of myself or my subjects against others.

In token whereof, I, Lieutenant Colonel Sir Percy Cox, Political Resident in the Persian Gulf, and I, Shaikh Abdullah bin Jasim bin Thani, have respectively signed and affixed our seal to this original document and four copies, thereof.

Dated 6th Moharram 1336, corresponding to 3rd November 1916.

ABDULLAH BIN JASIM
Chief of Qatar.

P. Z. COX, MajorGeneral,
Political Resident in the Persian Gulf.

CHELMSFORD,
Viceroy and Governor-General of India.

APPENDIX –X

GENEALOGICAL TABLE OF THE AL THANI (MAADHID) FAMILY OF QATAR

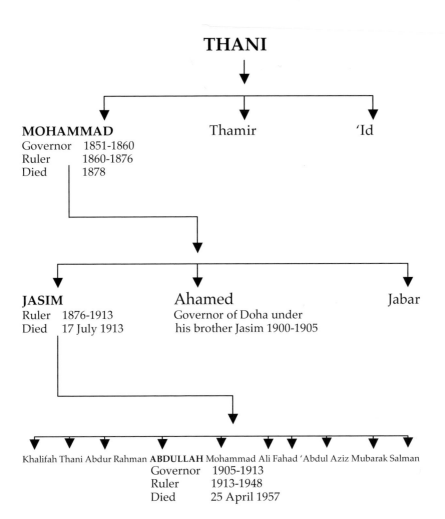

THANI

MOHAMMAD
Governor 1851-1860
Ruler 1860-1876
Died 1878

Thamir

'Id

JASIM
Ruler 1876-1913
Died 17 July 1913

Ahamed
Governor of Doha under
his brother Jasim 1900-1905

Jabar

Khalifah Thani Abdur Rahman **ABDULLAH** Mohammad Ali Fahad 'Abdul Aziz Mubarak Salman
Governor 1905-1913
Ruler 1913-1948
Died 25 April 1957

Source: Lorimer, <u>op. cit.</u>, Historical Part-3, Tables-Maps, pocket no. 8.

APPENDIX –XI

British Political Residents in the Persian Gulf, 1763-1917

Benjamin Jervis	Apr 1763 - Jul 1766
William Bowyear	Jul 1766 - Nov 1767
James Morley	Nov 1767 - Feb 1769
	(acting until Mar 1768)

Residency in abeyance between 1769 and 1775

John Beaumont	May 1775 - Jan 1781
Edward Galley	Jan 1781 - Jun 1787
Charles Watkins	Jun 1787 - [Oct] 1792
Nicholas Hankey Smith	1792 - Oct 1798
Mehdi Ali Khan	Oct 1798 - Jan 1803
Jonathan Henry Lovett	Apr 1803 - Jan 1804
Samuel Manesty	Feb 1804 - Mar 1804
Lt. William Bruce	Mar 1804 - May 1807(acting)
Nicholas Hankey Smith	May 1807 - Jul 1808
	(officially until March 1811)
Cpt Pasley	Jul 1808 - Sep 1808
Lt William Bruce	Sep 1808 - Jul 1822
	(acting until 1812)
Surgeon Andrew Jukes	Dec 1808 - May 1809(in charge)
Stephen Babington	April 1810(i/c)
Nicholas Hankey Smith	Sep 1810
Lt Robert Taylor	Apr 1812 - Nov 1812(i/c)
Asst-Surgeon James Orton	Aug 1813 - Apr 1814(i/c)
James Dow	Nov 1819 - Jan 1820(i/c)
	May 1820 - Jan 1822(i/c)
Asst-Surgeon John Tod	Jan 1822 - Mar 1822(i/c)
Jul-Dec 1822 Broker and Linguist in charge	
Cpt John MacLeod	Dec 1822 - Sep 1823
Cpt Henry Hardy	Oct 1823 - Dec 1823
Col Ephraim Gerrish Stannus	Dec 1823 - Jan 1827
Cpt David Wilson	Jan 1827 - Mar 1831
Dr. John MacNeil Dec 1830 - Dec 1831	
Lt Samuel Hennell	Mar 1831 - Jan 1832(i/c)
David Anderson Blane	Jan 1832 - Jun 1834
Lt. Samuel Hennell	Jun 1834 - c Jul 1835 (acting)
Surgeon Thomas Mackenzie	c Jul 1835
Major James Morrison	Oct 1835 - Oct 1837
Surgeon Thomas Mackenzie	Oct 1837 - Apr 1838(i/c)

Lt Samuel Hennell   [May] 1838 - Mar 1852
    Col Henry Dundas Robertson Jan 1842 - Apr 1842 (officiating)
    Lt Arnold Burrowes Kemball Apr 1843 - Dec 1843(offg)
Cpt Arnold Burrowes Kemball   Mar 1852 - Jul 1855
Cpt James Felix Jones   Oct 1855 - Apr 1862(offg to Jul 1856)
Cpt Herbert Frederick Disbrowe   Apr 1862 - Nov 1862(offg)
Lt-Col Lewis Pelly   Nov 1862 - Oct 1872
             (acting to Mar 1863; offg to Apr 1864)
Lt-Col Edward Charles Ross Oct 1872 - Mar 1891
                   (acting until 1877)
   Lt-Col William Francis Prideaux   May 1876   -   1877(acting)
   Lt-Col Samuel Barrett Miles   1885   -   Oct 1886(acting)
Lt-Col Adelbert Cecil Talbot   1891 - May 1893(offg to Sep 1891)
Cpt Stuart Hill Godfrey   May 1893 - Jun 1893(i/c)
Maj James Hayes Sadler   Jun 1893 - Jul 1893(acting)
James Adair Crawford   Jul 1893 - Dec 1893(i/c)
Maj James Hayes Sadler   Dec 1893 - Jan 1894(acting)
Col Frederick Alexander Wilson   Jan 1894 - Jun 1897
Lt-Col Malcolm John Meade   Jun 1897 - Apr 1900(offg to Mar 1898)
Lt-Col Charles Arnold Kemball   Apr 1900 - Apr 1904(acting)
Maj Percy Zachariah Cox   Apr 1904 - Dec 1913
         (offg to Oct 1905; substantive, temp to May 1909)
Maj Arthur Prescott Trevor   Aug 1909 - Mar 1910(i/c)
John Gordon Lorimer   Dec 1913 - Feb 1914
Cpt Richard Lockington Birdwood   Feb 1914 - Mar 1914(i/c)
Maj Stuart George Knox   Mar 1914 - Nov 1914(i/c)
Maj Percy Zachariah Cox   Nov 1914 - Oct 1920
      Maj Stuart George Knox   Jan 1915 - Apr 1915
                   (officer on special duty)
     Maj Arthur Prescott Trevor   Apr 1915 - Nov 1917

Source: Tuson, op. cit., p. 184.

APPENDIX –XII

British Political Agents in Bahrain, 1900-1918

John Calcott Gaskin	Feb 1900	-	Oct 1904
	(Political Assistant i/c)		
Cpt Francis Beville Prideaux	Oct 1904	-	May 1909
Cpt Charles Fraser Mackenzie	May 1909	-	Nov 1910
Maj Stuart George Knox	Nov 1910	-	Apr 1911
Cpt David Lockhart Robertson Lorimer	Apr 1911	-	Nov 1912
Maj Arthur Prescott Trevor	Nov 1912	-	May 1914
Cpt Terence Humphrey Keyes	May 1914	-	Mar 1916
Maj Hugh Stewart	Mar 1916	-	May 1916
J. M. Da Costa	May 1916	-	Jun 1916
	(Head Clerk in charge)		
Cpt Trenchard Craven Fowle	Jul 1916	-	Nov 1916(acting)
Cpt Percy Gordon Loch	Nov 1916	-	Feb 1918

Source: Tuson, op. cit., p. 185.

Bibliography

Unpublished Sources

1. Basbakanlik Osmanli Arsivi (BOA), Istanbul, Turkey:

    (a)   General correspondence Relating to Qatar
          (1884-1896)
          SD-2155/24.
          SD-2157/10.
          SD-2158/5/L.7, L.8, L.17.
          SD-2158/10/L.9, L.13, L.15, L.20, L.73, L.75.
          SD-2170/7/L.5.
    (b)   Inter-ministerial correspondence Relating to Qatar
          (1893-1905)
          Y.MTV-76/125, 133, 158.
          Y.MTV-170/4.
          Y.MTV-178/152.
          Y.MTV-272/137.
          Y.MTV-291/3.

2. Bombay Moharastra Archives:

    Secret and Political Department, Consultations
    (1810) Vol. 10/Folios - 1050 – 1055.

3. India Office Library and Records (IOR), London:

    (a)   Political and Secret Department Records
              Secret correspondence with India (1868 – 1871)
              L/P&S/5/261, 267, 268.
    (b)   Political and Secret Department Records
              Correspondence Relating to Areas outside India
              (1793 – 1874)
              L/P&S/9/15.
    (c)   Political and Secret Department Records
              Departmental Papers (1902 – 1918)
              L/P&S/10/4, 162, 384, 386.
    (d)   Political and Secret Department Records
              Annual Departmental Files (1912 – 1913)

L/P&S/11/46.
(e)   Political and Secret Department Records
        Political and Secret Memoranda (1840 – 1928)
        L/P&S/18/B133, B301.
(f)   Political and Secret Department Records
        Political and Secret Library (1916)
        L/P&S/20/C/158E.
(g)   Political Residency, Bushire (1778 – 1934))
        IOR: R/15/1/3, 19, 20, 22, 30, 38, 68, 69, 70, 71, 80, 84,
        85, 87, 91, 92, 94, 95, 99, 100, 101, 104, 105, 110, 111,
        114, 117, 125, 128, 130, 138, 141, 149, 179, 183, 203, 314,
        629.
(h)   Political Agency, Bahrain (1900 – 1920)
        IOR: R/15/2/25, 26, 27, 29, 30.
(i)   Political Agency, Kuwait (1909 – 1913)
        IOR: R/15/5/65, 68.

4.   The Public Record Office (PRO), London:

(a)   Admiralty File (1915)
        ADM 137/1152.
(b)   Cabinet Memorandum (1911)
        CAB 16/15.
(c)   Correspondence with Turkey (1887 – 1905)
        F.O. 78/5108, 5109, 5110, 5380.
(d)   Political Correspondence (1910 – 1934)
        F.O. 371/776, 1004, 1484, 1485, 17799.
(e)   Numerical Series (   )
        F.O. 881/9746, 9835, 9907.

5.   *Lam al-Shihab fi Sirat Mohammad bin Abd al-Wahab*, London:
        British Museum (MS. add) 23, 346.

Published Primary Sources

Aitchison, C. U. (Comp.). *A Collection of Treaties, Engagements and Sanads Relating to India and Neighbouring Countries*, vol. XI, Delhi: Manager of Publications, 1933.

Cook, Andrew S (Comp.). *Survey of the Shores and Islands of the Persian Gulf: 1820-1829*, vol. I, Oxford: Archive Editions, 1990.

Constable, Captain C. G. and Lieutenant A. W. Stiffe. *The Persian Gulf Pilot, including the Gulf of Oman*, London: Admiralty, 1864.

Hurewitz, J. C. *Diplomacy in the Near and Middle East: A Documentary Record 1535-1956*, vol. I, Gerrards Cross (UK): Archive Editions, 1987.

Lorimer, J. G. *Gazetteer of the Persian Gulf, Oman and Central Arabia*, Historical vols. IA and IB; Geographical and Statistical, vols. IIA and IIB, Historical Part3, Tables-Maps, Hantz (UK): Gregg International Publishers Ltd., 1970 (Reprint).

Low, Charles Rathbone. *History of the Indian Navy 1613-1863*, vol. II, London, Richard Bantley and Son, 1877.

Niebhur, Carsten. *Travels Through Arabia and Other Countries in the East*, 2 vols., Reading (UK): Garnet Publishing, 1994 (Reprint).

Saldanha, J. A. *The Persian Gulf Précis*, vols. I, IV and V, Gerrards Cross (UK): Archive Editions, 1986.

Thomas, R. Hughes (ed.). *Selections from the Records of the Bombay Government*, New Series, No. XXIV, 1856, New York: The Oleander Press, 1985.

The Persian Gulf Administration Reports: 1873-1920, vols. I-VII, Gerrards Cross (UK): Archive Editions, 1986.

The Persian Gulf Historical Summaries: 1907-1928, vol. I, Gerrards Cross (UK): Archive Editions, 1987.

Books: Arabic and English

Abu-Hakima, Ahmad Mustafa. *History of Eastern Arabia: 1750-1800*, London: Probsthain, 1988.

Abu Nab, Ibrahim. *A Story of State Building, Beirut*: Name of the Publisher is unknown, 1977.

Al-Abdulla, Yousof Ibrahim. *A Study of Qatari-British Relations 1914-1945*, Doha: Orient Publishing & Translation, 1982.

Al-Dabagh, Mustafa Murad. *Qatar: Madiha Wa Hadiruha*, Beirut: Dar al-Taliha, 1961.

Al-Mansour, Abdul Aziz. *Al-Tattawur al-Siyasi Li Qatar Fi al-Fatra Ma Bayan 1868-1916*, Kuwait: Chains Publications for Printing, Publishing and Distribution, 1980.

Al-Shibani, Mohammed Sharif. *Emarat Qatar al-Arabiya Bayn al Madhi Wa al Hadhir*, Beirut: Culture House, 1962.

Anscombe, Frederick F. *The Ottoman Gulf: The Creation of Kuwait, Saudi Arabia and Qatar*, New York: Columbia University Press, 1997.

Anthony, John Duke. *Historical and Cultural Dictionary of the Sultanate of Oman and the Eastern Arabia*, New Jersey: The Scarecrow Press Inc. 1976.

Busch, Briton Cooper. *Britain and the Persian Gulf, 1894-1914*, Berkeley and Los Angeles: University of California Press, 1967.

Davies, Charles E. *The Blood Red Arab Flag*, Exeter (UK): University of Exeter Press, 1997.

De Cardi, Beatrice (ed.). *Qatar Archaeological Report, Excavations 1973*, Oxford: Oxford University Press, 1978.

Dickson, H. R. P. *Kuwait and Her Neighbours*, London: George Allen & Unwin Ltd., 1956.

Farah, Talal Toufiq, *Protection and Politics in Bahrain 1869-1915*, Beirut: American University of Beirut, 1985.

Huede, Lt. William. *A Voyage up the Persian Gulf and a Journey Overland from India to England in 1817*, London: Longman, 1819.

Ibn Bashar, Shaikh Uthman bin Abdullah. *Aunwan al Majd fi Tariq Nejd*, Riyadh: Matba'at Daarath al Maalk Abdul Aziz, 1982.

Ibn Ghanam, Husain. *Kitab al-Ghazwat al-Bayaniyya Wal-Futahat al Rabbaniyya Wa Dhikhr al-Sabab Alkdhi hamal ala dhalitl*, vol. II, Bombay: Name of the Publisher is unknown, 1919.

Ibn Sanad, Uthman. *Sabik al-Azjad fi Akhbar Ahmad Najil Rizq al-Asad*, Bombay: Name of the Publisher is unknown, 1897.

Kelly, J. B. *Britain and the Persian Gulf: 1795-1880*, Oxford: At the Clarendon Press, 1968.

_____, *Eastern Arabian Frontiers*, London: Faber and Faber, 1964.

Kumar, Ravinder. *India and the Persian Gulf Region 1858-1907*, Bombay and New Delhi: Asia Publishing House, 1965.

Kursun, Zekeriya. *The Ottomans in Qatar: A History of Anglo-Ottoman Conflicts in the Persian Gulf*, Istanbul: The ISIS Press, 2002.

Moorhead, John. *In Defiance of the Elements: A Personal View of Qatar*, London: Quartet Books, 1977.

Palgrave, William Gifford. *Narrative of Year's Journey Through Central and Eastern Arabia 1862-63*, vol. II, London and Cambridge: Macmillan & Co. 1865.

Pearson, J. D. (ed.). *A Guide to Manuscripts and Documents in the British Isles Relating to the Middle East and North Africa*, Oxford: Oxford University Press, 1980.

Rahman, H. *The Making of the Gulf War: Origins of Kuwait's Long-Standing Territorial Dispute with Iraq*, Reading (UK): Ithaca Press, 1997.

Slot, B. J. *The Origins of Kuwait*, Kuwait: Centre for Research and Studies on Kuwait, 1998.

Tuson, Penelope. *The Records of the British Residency and Agencies in the Persian Gulf*, London: India Office Library and Records, 1979.

Vine, Peter and Paula Casey. *The Heritage of Qatar*, London: IMMEL Publishing Ltd., 1992.

Wilkinson, John C. *Arabia's Frontiers: The Story of Britain's Boundary Drawing in the Desert*, London and New York: I. B. Tauris & Co. Ltd., 1991.

Wilson, Sir Arnold T. *The Persian Gulf: An Historical Sketch from the Earliest Times to the Beginning of the Twentieth Century*, London: George Allen & Unwin Ltd., 1954.

Yapp, M. E. *The Making of the Modern Near East: 1792-1923*, London and New York: Longman, 1987.

Zahlan, Rosemarie Said. *The Creation of Qatar*, London: Croom Helm Ltd., 1979.

Index